# THE GLOBAL TRANSFORMATION OF TIME

# THE GLOBAL TRANSFORMATION OF TIME

## *1870–1950*

VANESSA OGLE

Harvard University Press
*Cambridge, Massachusetts*
*London, England*
2015

Printed in the United States of America

First printing

*Library of Congress Cataloging-in-Publication Data*

Ogle, Vanessa.
The global transformation of time : 1870–1950 / Vanessa Ogle.
pages cm
Includes bibliographical references and index.
ISBN 978-0-674-28614-6 (hardcover : alkaline paper)
1. Time—Systems and standards—History.
2. Time—Sociological aspects.
3. Globalization. I. Title.
QB223.O35 2015
389'.17—dc23 2015006973

# Contents

# THE GLOBAL TRANSFORMATION OF TIME

# Introduction

IN THE SPRING of 1891, Count Helmuth von Moltke rose to speak in the German parliament on the adoption of "uniform time." His ninety years of age barely showed, for contemporaries later credited his energetic and inspiring speech with swaying an undecided audience toward the support of the proposed change of time. After several years of discussion among ministries, railway officials, legislators, and the public, Germany was finally ready to consider the nationwide implementation of a common mean time. It was less clear whether such a new time would apply only for internal use in telegraphy, railways, and perhaps select government offices, or whether the new mean time would be extended into all aspects of civil life. Even Moltke himself urged caution, given the magnitude of changes he and others understood to be part of such a switch. Moltke did not live to enjoy the fruits of his oratory labor; the grand old man of German military strategy passed away barely a month after his performance in the Reichstag. But in the fall of 1892 and again in the winter of 1893, the German parliament discussed a bill that proposed to make the time one hour faster than Greenwich, United Kingdom, the new mean time for all of the German Empire. The bill passed and became law on April 1, 1893, for both internal administrative and external purposes of daily life.[1]

Helmuth von Moltke was addressing a subject that was arguably one of the most important social, political, and cultural transformations wrought by a long nineteenth century: the emergence of modern times. In an arduous and drawn-out process, local times were abolished in favor of time zones and countrywide mean times; the Gregorian calendar spread to parts of the non-Western world; time was eventually severed from natural and agricultural rhythms and instead assumed more abstract qualities, a grid to be grafted onto natural rhythms; time was increasingly linked up with occupational notions—work time, leisure time, recreational time, time for acquiring useful knowledge. In many parts of the world, these transformations for almost a century resulted in an even greater variety of times, as religious and other times and local calendars continued to be used alongside new times. Hybridized and again transformed by changing patterns of communication and occupation in the postindustrial society, these modern times are still with us today.

Moltke was no unknown quantity. For some thirty years, he had headed first the Prussian and then the German general staff. His military prowess was widely regarded to be a leading cause behind Prussia's victories in three "wars of unification," conflicts that paved the way for the establishment of a German nation-state in 1871. In the parliamentary debate of 1891, Helmuth von Moltke's sleight of hand was therefore to play the card of the military leader. Perhaps Moltke's most distinct feature as a nineteenth-century strategist had been his knowledge and utilization of railways in the service of war. Railroads and supply lines had already played a role in the 1866 war with Austria, but the real test came in the 1870–1871 conflict with France. Here, Moltke deftly organized troop movements with the help of railways, not least thanks to a meticulously orchestrated system of schedules and connections.[2] When the old man made his case for uniform time before the Reichstag, he stressed the indispensable nature of well-timed railways for the defense of the German fatherland. Even in peacetime, he warned, the annual mobilization drill depended on the smooth and flawless movement of troops with the help of railways. For all these reasons, Moltke contended, the adoption of a uniform German mean time was a matter of national security, as it stood to greatly improve a key element in defense. Opposing it meant rendering Germany vulnerable.

Moltke advanced a second argument as to why Germany needed uniform time. Currently, a total of five times were followed in the country: Berlin, Ludwigshafen, Karlsruhe, Stuttgart, and Munich time. Such a multitude of times was, according to Moltke, "ruins that had been kept standing from the time of German fracture, which now that we have become a Reich [the German Kaiserreich, imperial Germany], would rightly have to be abolished." The general was referring to the political situation prior to 1871, when Germany consisted of numerous small, often tiny territorial states loosely joined together in the German Confederation. Doing away with the regionalism inherent in keeping five different times was as much an act of national security as of nation-building.[3]

Moltke's national reasoning notwithstanding, once these plans took the form of a bill placed before the German parliament, they were interspersed with a very different line of argument. The explanatory statement accompanying the law painted the picture of a world that was becoming smaller, in which exchange was proliferating and distances were diminished, and in which people, goods, and ideas incessantly crossed national borders. "As long as life moved overwhelmingly in narrow circles," the draft bill stated, "as long as traffic from one place to another was slow and the exactitude of determining time only low, local times . . . were sufficient for every single locality." Now, however, "there is an increasingly widespread realization that refined, space-and-time commanding means of transportation [and by implication, communication] mandate the creation of a uniform definition of time for greater territorial areas."[4] The statement was a nod to the proposal for adopting a single meridian on which to base a worldwide system of hour-wide time zones, voiced most prominently at the 1884 Prime Meridian Conference in Washington, DC.[5] That system, and slightly later, its less well-known twin project of introducing a uniform "world calendar," obeyed the logic of an almost utopian vision of universal time, of connectivity and growing integration. Evidence of such evocations of an interconnected world is ample and ranges from journalistic essays, pamphlets on time unification, tracts penned by scientists, and statements by legislators, administrators, and watchmakers. What at first blush might appear as mere rhetoric was too deeply engrained a mind-set and worldview to be dismissed as window dressing.

The drafters of the German bill were right: the latter decades of the nineteenth century saw dramatically enhanced interactions on a global scale, drawing people and regions into networks of production and exchange, into imperial and colonial careers and livelihoods, while disrupting autonomous, regional economic and political systems in non-Western societies. As early as 1848, Marx and Engels put a socioeconomic spin on interconnectedness when writing that "the need of a constantly expanding market for its products chases the bourgeoisie over the whole surface of the globe. It must nestle everywhere, settle everywhere, establish connections everywhere." Telegraphs, railways, and steamships furnished new technological and communicative underpinnings for such interactions. The nineteenth-century world, to use a present-day term, was rapidly becoming more globalized.[6] Globalization is understood here in its broadest sense as the process of forging sustained political, economic, social, and cultural interconnections, exchanges, and dependencies between world regions and states. In the second half of the nineteenth century, in some of its features, globalization looked not unlike it does today. Writing about the decades around 1900 as a period of globalization helps make the bygone world of the nineteenth century more familiar while bringing distinctions between the two epochs into focus. At the same time, such a perspective raises doubts about the often-claimed novelty and unprecedented scale and scope globalization is said to have taken on today.[7]

In his reading of universal time as German national time, Helmuth von Moltke was barely alone. The apparent tension between universalizing ideas about uniform time, the spread and flow of such ideas, and a nationalist and sometimes regionalist framework of implementing such ideas was not lost on contemporary observers. Robert Schram, an Austrian geodesist reflecting on the adoption of nationwide mean times, observed as much. "Our century displays strange opposites," he wrote. "While on the one hand mankind is seeking to separate itself into closely sealed-off groups of nations and little nations [nationhen] and to erect dividing walls among them, on the other hand, [mankind] feels to unprecedented degrees the need for commonalities in everything that has to do with trade, industry, and technology."[8] Uniform time was one such commonality.

• • • •

AROUND 1900, contemporary observers were aware of just how much interconnectedness was redefining time, space, and social relations. They pictured a world that was turning into a global village. What sets the second half of the nineteenth century apart from earlier examples and periods of dense connections and interactions is this apprehension of globality, the realization of and reflection on the fact that the world was interconnected. David Harvey coined "time-space-compression" as an analytical concept for capturing such transformations in hindsight, and others have followed in his vein. But such "connectivity talk" and self-conscious contemplations on globality were first expressed during the nineteenth century itself. Self-reflective globalization was a mentality that shaped how people responded to and sought to steer globalization.[9] At the same time, Moltke and his fellow nineteenth-century Europeans and Americans accurately thought about the globe as citizens and subjects of a world of competing nation-states, where national identities and national belonging increasingly took center stage. A global history of timing practices, of ideas about reforming time, brings to the fore the dynamics at work in an interconnected world: the spread of ideas from Europe and North America to other parts of the world and their implementation initially as national and regional times, but also their simultaneous articulation in different parts of the world without direct channels of transmission. The origins of what would become universal time only in the aggregate were national, local, and sometimes regional. The story of time reform thus raises perhaps the most challenging analytical question that every historian working on the global nineteenth and early twentieth centuries confronts: the relationship between increasing integration and connections between world regions and states on the one hand and the simultaneous rise of nationalism and powerful state apparatuses on the other.[10] Such a global and international perspective, moreover, places European developments next to other histories. It constitutes a proposal to write European history "from the outside in" by examining how different European countries have been shaped by their interactions with and perceptions of other world regions, and by the apprehension of an increasingly globalized world.

When Moltke and others looked around in the nineteenth-century version of the global village, what struck them was how fiercely competitive this shrinking world had become. The overcoming of distance seemed to signal not just the facilitation of travel and communication, but general encroachment, making peaceful coexistence in diminished space evermore difficult to imagine. In Europe, the continent was struggling to accommodate the rise of a new, ambitious, and increasingly aggressive great power and economic might: Germany. Outside of Europe, a territorial grab for the last remaining uncolonized swaths of land and the opening of new markets for acquiring and selling raw materials, commodities, and manufactures was in full swing, with even Japan and the United States now marking their own realms of influence in Asia and the Western Hemisphere. This impression of zero-sum competition was not even restricted to great power politics. It permeated the sciences and generally different academic disciplines, for expertise and knowledge could translate into industrial, military, and cultural "progress" and allow one nation to overpower others. From the point of view of those residing outside of Europe and North America, the prospects were no less troubling. Chinese, Ottoman, Egyptian, and Levantine observers were alarmed at the sight of non-Western societies falling under Euro-American colonial and imperial rule in what appeared to be an irreversible and inexorable process. In this interconnected world of cutthroat competition, both non-Western societies and Moltke and his contemporaries in Europe and North America often sought to bend the flow of goods, people, and ideas about time to serve the improvement of their nations and the strengthening of their own societies and states.[11]

The single most important intellectual device by which nineteenth-century observers gauged an interconnected and competitive world was comparison. Whether in Berlin or Beirut, London or Bombay, contemporaries were obsessed with establishing differences and similarities between other societies and nations and their own. They compared everything from the level of civilizational progress to economic strength, demographic developments, the shapes of human skulls, scientific achievements, and various ways of measuring clock time, designing calendars, and managing social time. Evolutionary thought, so intriguing in the second half of the nineteenth century even in parts of the non-Western world, lent itself particularly well to diachronic comparisons of different stages of biological, social, and cultural development and refinement and became an underlying epistemic tool in many fields of knowledge production.[12]

Comparison was facilitated by the structures put in place by nineteenth-century globalization. Cheapening print products disseminated news and information about the achievements and shortcomings of foreign societies and nations with more ease and at a lower price than in previous ages, opening a window onto the world for a growing number of people. What is more, an expanding group of individuals experienced foreign societies at first hand, through trade, travel, migration, and in the service of imperial and colonial endeavors. Comparison translated between the global panorama of economic, political, and cultural trends and the level of local experiences. The German philosopher Friedrich Nietzsche expressed this relationship in an aphorism in 1878:

> *Age of Comparison.* The less men are fettered by tradition, the greater is the fermentation of motivations within them, and the greater in consequence their outward restlessness, their mingling together with one another, the polyphony of their endeavors. Who is there who now still feels a strong compulsion to attach himself and his posterity to a particular place? Who is there who still feels any strong attachment at all? Just as in the arts all the genres are imitated side by side, so are all the stages and genres of morality, custom, culture.—Such an age acquires its significance through the fact that in it the various different philosophies of life customs, cultures can be compared and experienced side by side; which in earlier ages, when, just as all artistic genres were attached to a particular place and time, so every culture still enjoyed only a localized domination, was not possible.[13]

Take Edmond Demolins, a French pedagogue who published "Anglo-Saxon Superiority—To What Is It Due?" in 1897, as one popular example of a comparative mind that was born out of the experience of a globalizing world. The tract, infused with racial theories, compared British and French education systems and lambasted the French system for contributing to the demographic decline of the nation as well as its financial woes. Translated into Arabic, Bulgarian, English, Farsi, Japanese, Polish, Russian, and Spanish (in Spain and Peru), the title was similarly popular among others curious about the secrets of national success.[14]

The problem with comparison was that globalization and the supposed overcoming of time and space made contemporaries realize not how similar but how heterogeneous the world was—to the degree of being incomparable. In this world, it was time that served to measure and establish difference. Time was therefore always more than just one other category or standard that was internationalized and universalized in these decades, such as law, the metric system, or the women's movement. Time is what made the global imagination possible in the first place. Time was ubiquitous: in the second half of the nineteenth century, time became an object of display and inquiry in a staggeringly wide array of fields. Some of these areas have found their historian in Stephen Kern, who devoted a fascinating book to this "culture" of time and space, allowing him to view intellectual and artistic expressions as different as European and occasionally American literature, painting, philosophy, and geography, and others, through the prism of various qualities of time and space.[15]

For one, history as a discipline emancipated itself from philology and other fields and became an integral part of the university as it emerged in its modern form. Other sciences such as archaeology and geology were similarly investigating historical time. As is well known, these (and many other) disciplines first and foremost served the canonization and often invention of national histories for the nation-state.[16] But historical times also helped create relations of difference by plotting the histories of nations and peoples onto a grid of universal, evolutionary time, thus situating national histories in a world of time. In his famous interpretation of nationalism, Benedict Anderson has emphasized the importance of the idea of a nation as moving "steadily down (or up) history," analogous to the notion of an "organism moving calendrically through homogenous, empty time," for imagining the community of the nation. But what Anderson did not see, what is perhaps the most helpful achievement of transnational history as yet to have illustrated, is how nation-states and nationalism were shaped as much from the outside-in as by domestic populations coming to imagine themselves as belonging to a nation. Borders, imaginary and physical, were made when crossed and defined by what lay beyond them as much as within them.[17] The nation was imagined not only as a national community but as a nation that formed part of a global community of societies and other nations, all positioned in historical time. In consequence, those non-Western societies either deemed "peoples without history" or reduced to an earlier stage of civilizational and evolutionary development were destined to be the subject of anthropology rather than history in the nineteenth-century academic division of labor. As such, Euro-American historians, anthropologists, and ethnologists denied them "coevalness" as Johannes Fabian has argued in his work on time and anthropology.[18] Such a temporalization of difference and different temporalities were of utmost importance in nineteenth-century globalization. Time, or the absence thereof, thus became a measure for comparing different levels of evolution, historical development, and positionality on a global scale.

Social time, too, served as a location device of sorts. In the second half of the nineteenth century, workers and employers alike struggled over the meaning and division of social time, of work time and leisure, and the appropriate proportion between them. Assigning certain divisions of time-specific tasks such as "work," "leisure," and "rest" gave rise to a heightened awareness for efficient time management among these collectives. Citizens and subjects were moreover exposed to time discipline as practiced in schools, factories, the military, and other "complete" institutions, as Michel Foucault has termed them.[19] As part of nineteenth-century globalization, classifications among individuals, collectives, and their functionally differentiated times were soon applied not only with regard to domestic socioeconomic groups but extended to capture the working habits and time management skills of entire peoples and societies or, more likely, races. In the eyes of Western missionaries and businessmen alike, non-Westerners were collectively indicted as work-shy and incapable of proper time management. In a world of imperial competition, these stigmatizations were occasionally internalized by reformers especially in areas that had not yet fallen under European rule but stood at risk of succumbing to Western superiority. Arab intellectuals in the Levant urged their contemporaries to improve their time management in order for the "Eastern" civilization to catch up with Europeans. The nineteenth-century preoccupation with progress and modernity was built on similar notions of linear historical and naturalized, evolutionary time. Marching in sync with progress moreover mandated the modern management and deployment of time. In the non-Western world in particular, mechanical clocks thus became status symbols, displaying the owner's keeping up with modern times.[20] The aspiration to be modern, so captivating to scores of intellectuals and reformers far beyond Europe and North America, thus became another tool for global positioning.

The rational and efficient time associated with progress also pervaded the swelling bureaucracies and administrations of nineteenth-century states. Fine-tuning the rational time of trains and telegraphs, of schedules and timetables, and streamlining their administration became one of the major preoccupations of nation-states eager to cast a technocratic grid of time over national territory. But for national times to be rational and efficient, they had to be integrated with other national times, initially mainly neighboring ones but increasingly so, a spate of other times originating far beyond a state's immediate borders. Even when implementing nationwide mean times, nineteenth-century contemporaries imagined the place of the nation in regional time, if not always immediately in world time.

Calendars and the systems of dating they entailed were another method for establishing temporal differences between societies and nations by anchoring them in particular religious as well as universal histories. The second half of the nineteenth century was a time of calendar reforms. The Gregorian calendar was introduced in a number of non-Western countries. By the end of the century, several religious groups discussed reforming their respective calendars, and Euro-American businessmen demanded the adoption of a uniform "World Calendar." Comparing calendar times and devising conversion formulas for navigating different calendars and histories became a major preoccupation among economists, calendar reformers, and curious laymen alike.

While history and historical thinking, calendar reform, time management and efficiency, and progressive time found many followers in the non-Western world, it was the peculiarity of Europeans and Americans to devise all-encompassing, universal plans for time reform. Certain differences appeared too incommensurable to be mapped out via simple comparisons, and more uniformity would bring greater comparability. According to Marx and Engels, the propagation of uniformity was the prerogative of one class in particular, and was not even limited to time. In the words of their famous *Manifesto,* "The bourgeoisie, by the rapid improvement of all instruments of production, by the immensely facilitated means of communication, draws all, even the most barbarian, nations into civilization. . . . In one word, it creates a world after its own image."[21] The promotion of a system of twenty-four uniform time zones each one hour wide by European and American astronomers, diplomats, and railway men was the most ambitious such plan. Everywhere in the world, local solar times would be abolished to make room for new mean times that, given their uniformity, would be easily comparable and calculable. Similarly, a uniform world calendar would abolish the countless local systems of dating and arranging the seasonal year and replace them with one uniform year, one single dating system for all. In these very same years, Muslim learned men from around the Eastern Mediterranean envisioned a similarly universal time based in astronomy—as Islamic timekeeping had relied on astronomy for centuries even before Europeans did—only in their case, a universal Islamic calendar would apply to all Muslims but refrain from homogenizing all times by superimposing Islamic time onto other, existing times.

• • • •

IN THE SECOND HALF of the nineteenth century, these variegated preoccupations with time—ranging from mean times to clocks and watches, social time, time management, and calendar time—made for a strikingly simultaneous emergence of "time talk" around the globe. Among these varieties of time, one set of properties in particular has received considerable attention among historians. Historically, time measurement is widely understood to have evolved from depending on natural, "concrete" timekeepers such as the sun and the seasons to what is commonly understood as "abstract" time, measured first by mechanical and eventually by atomic clocks. In this vein, Moishe Postone has juxtaposed "concrete" and "abstract" time in his reworking of Marxist thought and characterization of social relations under capitalism. Concrete time is a function of events and refers to particular tasks or processes often related to agricultural labor or to saying a paternoster. Reckoning concrete time depends on natural events such as days, lunar cycles, and seasons rather than on the succession of uniform temporal units. Historically, in various cultures including European ones, periods of daylight and darkness were divided into twelve uneven hours each, varying in length with the seasons. In the fourteenth century, what Postone and others refer to as "abstract time" emerged in the urban centers of northwestern Europe. Postone's abstract time is "uniform, continuous, homogenous, 'empty' time, unrelated to events." By the seventeenth century, abstract time "was well on its way to becoming socially hegemonic." Time was now a measure of, and eventually a norm for, activity.[22] E. P. Thompson's famous article "Time, Work Discipline, and Industrial Capitalism," almost fifty years old at this writing, had made related arguments by contrasting "task orientation" and the rhythms of nature with "time-sense" and the internalization of factory discipline and the valuation of the empty time of uniform working hours.[23]

Contrary to this narrative (which Postone mostly advances as a historical background and not primarily as an argument), the story of global time reform reveals that abstract time was not a given entity or quantity that was gradually uncovered over centuries and eventually standardized in a worldwide process. Presumably universal, uniform, and abstract time was actively made and remade, invented perhaps in a drawn-out, arduous, and laborious process that took much longer to achieve than commonly assumed, and that far from being an independent variable, relied on various material, political, and cultural and social realities. For many, even educated people, understanding time as abstract therefore posed insurmountable imaginary difficulties as late as the first two decades of the twentieth century. Neither the distribution and display of the new, uniform time on public and personal clocks nor industrial capitalism had by this time generated a world of homogenous, abstract time.

Time is so broad a topic that numerous books have been written about many of its facets, from antiquity through the Middle Ages to the present. All societies have, in some form or another, felt the need to mark time. They deployed a wide array of tools and instruments, some simple, some highly sophisticated, to satisfy that desire. Accordingly varied is the background of those writing on time: historians of different eras, historians of science and technology, anthropologists, sociologists, and philosophers, to name a few. Historians often associate the topic of "time" with the classic stories of medieval and early modern clock times told by David Landes, Carlo Cipolla, and Jacques Le Goff. Landes's book tells at least three stories: why Europeans invented the mechanical clock in the Middle Ages; why the Chinese failed to do so; and finally, the story of the emergence of the modern watch industry in eighteenth-century Europe and in the early nineteenth-century United States. Landes's account oozes a triumphalism celebrating the achievements of European genius over the shortcomings of others, especially the Chinese "civilization." It cannot seriously stand as a model for writing the history of time in a cross-cultural and comparative perspective. Carlo Cipolla's short book charts the proliferation of mechanical clocks in fourteenth-century Europe and their increasing sophistication in the following three centuries. Among the arguments about medieval and early modern time, Jacques Le Goff's intervention has proved most insightful and lasting. In a short article, Le Goff proposed that the emergence of a commercial market society in fourteenth- and fifteenth-century Europe was accompanied by a growing utility for marking time, hence the proliferation of work bells, tower clocks, and other visible and audible signals denoting various times. Ever since, the emergence of capitalist society has been closely tied to the proliferation of clocks and the honing of time consciousness. How widely such practices and notions of time spread in the following centuries, and how closely capitalism was really enabled by or dependent on these developments, is a different question that Le Goff does not address.[24]

Beyond the Middle Ages and early modern times, time has captured the attention of historians of the nineteenth century, and in particular, historians of the United States and French science. Accounts exist on the commercialization of time distribution services and the widespread mistrust toward new times, the spread of public clocks, and the American involvement with the international movement of scientists and observatory men lobbying to adopt uniform time zones. Due to the impact of scientific knowledge and communication technologies on time, the story of time reform has often been told by historians of science and technology. Such accounts analyze the challenges of time measurement that had vexed astronomers, mapmakers, navigators, clockmakers, and the rulers in whose service they stood at least since the age of explorations. Seafarers in particular craved chronometers accurate and stable enough to be taken aboard a ship while maintaining correct time. Countries such as England and France, evermore dependent on navigation for reasons of empire and commerce, endowed prizes generously rewarding anyone solving the problem of finding the longitude at sea. The reason was that time was inextricably tied to space. A full rotation of the earth, completed in twenty-four hours, equals 360 degrees. Hence, one hour can be expressed as fifteen degrees of longitude. The difference in longitude between two places is the same as the time difference between these two points. If a navigator knew the time difference between a fixed point, for instance at the point of origin of a journey, and apparent local time on the ship at a given point, he was able to derive a ship's location relative to the fixed point.[25]

With the help of astronomical tables, finding apparent local time at sea was possible. The real problem was to know what time it was in just that moment at the fixed reference point back home. Beginning in the mid-eighteenth century, some of the better maritime chronometers and pendulum clocks, set to local time at the origin of a journey, were finally robust enough to keep the original time with sufficient accuracy while on a ship journey.[26] Britain's Board of Longitude was a leading sponsor behind improvements in timekeeping and charts and maps. The resulting advancements in cartography established British mapmakers as leaders in the field. Cartographers in other countries began basing their own maps on the meridian of Greenwich, although a variety of meridians remained in use. In the nineteenth century, scientific congresses in different fields began to address the plurality of meridians and the absence of uniformity in worldwide timekeeping as a problem.[27]

Based on the archival papers of the scientists and scientific institutions involved, this history of time reform appears as one featuring influential astronomers, observatory scientists, geodesists, surveyors, mathematicians, and railway men who met at international conferences, exchanged ideas about worldwide timekeeping, chose a prime meridian, and devised time zones, most notably so at the 1884 conference in Washington, DC.[28] While these stories are fascinating accounts of networks of learning and the environments that inspired scientists to think differently about time, they should be expanded in various ways. Time reform was a global phenomenon at the turn of the twentieth century. The impetus behind propagating universal time and the logic that subsequently shaped time reform, the movement of ideas about time, their inflections, and the simultaneous articulation of similar concepts of time in various non-Western world regions cannot be captured in a national, American or French, perspective.

What is more, a history of time reform that focuses too squarely on experts, scientists, and diplomats between international conferences and institutions would risk misinterpreting the nature of time reform. Euro-

American experts on time surely imagined their plans for time reform to extend to the world beyond their own. Their timekeeping schemes encompassed the entire globe, and reformed time was "universal" and thus would not be confined to specific societies and world regions. But upon closer scrutiny, these academics, diplomats, and scientists brandished a rather limited notion of the globe. The "world" in world standard time, world calendar, world language (Esperanto), or world postal union above all stood for Europe and North America. Those deemed active participants in the spread and adoption of global reform plans were, with few exceptions, the representatives of sovereign nation-states in the Western world.[29] In such instances where the crowd of reformers gathering at international conferences or setting up international organizations did actively ponder the place of the non-Western world in various reform projects, their mental maps displayed other biases, reflecting spheres of interest and power in the age of empire: North Americans habitually looked to the Western Hemisphere and Latin America. Imperial-minded Britons turned to Australia or India as two different cornerstones of the variegated worlds of the British Empire.

The archives of Euro-American experts would perhaps point us to debates over countrywide mean times adopted in Latin America, as US officials reflected on the strategic importance of that continent to American interests, or to quarrels over time reform in India, as British scientific societies egged on government officials in London to endow the "crown jewel" of the British Empire with a more "civilized" mean time. These archives would almost certainly miss the movement and subsequent appropriation and transformation of ideas about time among reformers in the Levant, and the simultaneous articulation of calendar reform plans among Chinese, Ottoman, and Arab-Muslim contemporaries. On the mental world map of Euro-American globality around 1900, entire continents thus remained colored in white. This is not to say, for instance, that no one in Britain thought of Africa or that no one in the United States pondered Asia, as obviously the "scramble" for the conquest of Africa and the orientation of the American West to Asia and the Pacific in this period contradict such assumptions. But when internationalists talked about global governance and the implementation of reform schemes around the world, certain societies appeared more readily in conversations while others who might have simultaneously engaged in related discussions dropped off the map as being either unable—not yet ready to participate in such debates—or unworthy of association. Telling global histories of time reform (and other topics global in nature) is not a matter of all-inclusiveness and completeness. But the risks of reproducing the hierarchies of knowledge and power that emerged as part of the very process that is under study here are real. Global histories of time reform and other continent-spanning developments therefore have to strategically counteract these distortions. Such histories cannot simply follow the official mind of expertise on time reform. One way to counterpoise the blind spots of the age of empire is to follow the movement and articulation of ideas about time or any other matter to a variety of political units and scales that together made up the world of 1900—broadly conceived, centralized nation-states, overseas empires, multiethnic, not fully colonized land empires, cities, and international organizations, to name some of the most central typologies. Such a conceptualization of what "global" and "international" can mean to the historian of modern Europe and beyond offers a perspective that brings into focus the positionality of whoever was thinking the universal—that is, the social and political conditions under which historical actors thought and conceived of "the world" in different places and societies. Time reform tasted different from the point of view of the western European nation-states Germany, France, and Britain; recent acquisitions in colonial Asia and Africa; the long-standing British possession in India; spheres of informal European interests in Latin America; urban Beirut at the periphery of the struggling Ottoman Empire; and international organizations such as the League of Nations.

Just as important, government archives as opposed to the archives of experts, observatories, and international organizations paint a very different picture of the implementation of laws about clock times and calendars. In this perspective, the 1884 conference in Washington, DC, in which experts were so invested, appears not as an endpoint but rather as a modest and soon completely forgotten beginning of time reform, remaining largely without influence. To the adoption of mean times, the 1884 conference was almost meaningless as the process of time unification dragged on until the 1930s and 1940s and was all but unstable, precarious, and full of malfunctioning, even in Europe. Outside of Europe and North America there was no system of time zones at all, often not even a stable landscape of mean times, prior to the middle decades of the twentieth century, about fifty to sixty years later than assumed in extant scholarship. What scientists had in mind in devising a scheme for worldwide time zones was, after all, an incredibly ambitious project. The practical difficulties of ensuring that a law about time was actually followed, and the logistic challenge of orchestrating the comprehensive distribution of accurate time, were serious tasks that took decades to accomplish. Crucially, once time reform was discussed outside the circles of scientists and observatory operators, it was tremendously difficult even for lawmakers and at least moderately educated bureaucrats to imagine time as abstract and empty, untethered from the rhythms of nature and the movements of the sun. Discussions about reforming clock times and calendars, the arguments for and against such steps, reveal much more about contemporaries' ideas about time than has been acknowledged.

A more variegated archival basis moreover points to the multidimensional meaning of reforming and transforming "time" around the turn of the twentieth century. There was a social, political, and cultural side to time reform that was much more central to the story than international conferences and observatories and their deliberations about standard time suggest. Scientists, astronomers, and railway men may have focused their energies on accuracy and measurement, whether for the purpose of longitude determination or timetable design. But government officials and legions of journalists, publicists, and self-declared "experts" understood time to have political implications for the social use and experience of time in everyday and work life. Beginning in the first decade of the twentieth century, government officials and other experts therefore grew increasingly concerned with the deviation of mean time from solar time and the most prudent use of daylight.[30] Administrators in Europe and the United States started worrying about saving daylight and scolded what they perceived as a "waste" of light especially during the summer months. More explicitly than in previous discussions about mean times, in their view, time was now infused with social ideas and meanings. The adoption of summer or daylight saving time was debated as an instrument to change people's behavior, to encourage certain activities and usages of time and to discourage others. In an age when the working class was expanding, when more democratic political participation became a reality and to some, a threat, daylight saving time was hailed as an expedient for forging good citizens.

The state was an ambivalent source of new times. Beyond Europe and North America, colonial governments did not always see an immediate need to expand the use of uniform, hour-wide time zones. Societies outside of Europe and North America therefore experienced the transformation of time not merely through a state or government. In instances where mean times were introduced early, as was the case in British India, they had the potential to incite protests and refusal precisely because these new times emanated from the colonial state. In other cases, where the existing state refrained from or was simply too thinly spread for tackling the logistics of time unification, a globalizing world, with its reconfigurations of time and space, inspired middle classes to reflect on the changing nature of time and space out of their own interests in the comparison of times. In the Arab provinces of the Ottoman Empire, contemporaries did not even have to turn abroad for comparisons. European and American consuls, missionaries, and merchants were landing on Eastern Mediterranean shores in growing numbers, founding schools, universities, and their own courts of law. In this context, comparing "Arab" and "Islamic" times to Euro-American ones came naturally to reform-minded observers.

The widening presence of Europeans and Americans in the late Ottoman Levant rattled local journalists and intellectuals not least because simultaneously, Europeans formally colonized an entire continent not all that far away. Anxious, Arab and Muslim thinkers embarked on a mission of self-strengthening and self-improvement of the "East," as they understood it. Part of this movement was to instill more efficient management of time into their fellow Arabs. Comparing Euro-American approaches to the use and spending of social time with the habits prevalent in their own society suggested that Arabs needed to do better if they were to stand up against foreign influence and possibly colonial rule. As a remedy, intellectuals and journalists advocated what amounted to an amalgamation of much older Euro-American notions of "time is money" and even more seasoned Islamic and pre-Islamic philosophies about time as fate. Local reformers chose essays and articles in the flourishing Arab-language press as the primary medium for communicating their reflections but also, occasionally, poems about telegraphy, clocks, and time.[31]

Not everyone endorsed modern technology and the times it wrought wholeheartedly. At the beginning of the twentieth century, Muslim legal experts and learned men engaged in a heady debate about whether it was in accordance with Islamic law to use a telegram to report the sighting of the new moon, and whether it was permissible to act on such a telegraphic message and declare the beginning and end of the fasting month of Ramadan. Such controversies ultimately raised the question of just how universal Islamic calendar time could be, given that Muslims lived all over the world. In a world connected by print and telegraphy, Muslims from Southeast Asia to North Africa suddenly noticed the often starkly divergent timing practices for determining the Islamic lunar calendar. Like their Euro-American contemporaries, Muslim legal scholars parsed the scriptures for evidence on uniform calendar determination and ultimately came up with the concept of a universal Islamic calendar.

Calendar reform was a widely discussed matter in the Western and non-Western world alike. In Europe and North America in the 1870s and 1880s, scientists and astronomers focused their energies on clock times. It has been overlooked that soon their interests turned to calendars as well. Reformers, legislators, and administrators understood "time" to include not just clock times and the adoption of mean times, but also calendars. In his efforts to reorganize German clock times, Helmuth von Moltke was joined by countless calendar reformers and their attempts to make calendar time more uniform—Cesare Tondini de Quarenghi, a Barnabite monk who in the 1890s peregrinated between Bologna, Paris, and St. Petersburg, working toward the adoption of the Gregorian calendar in Orthodox countries; George Eastman, founder and owner of Kodak, who advocated the adoption of a thirteen-month calendar in the 1920s and 1930s; and Ahmad Shakir, a Cairene scholar of the Qur'an and a jurist who in 1939 proposed a universal Islamic calendar.[32]

In European government archives, correspondence on calendar reform is filed away in the same ledgers that conserve the documents on the adoption of mean times. Contemporaries clearly saw both calendars and clock times as pertaining to the broader issue of time unification, yet no account of calendar reform by modern-day historians exists. Euro-American calendar reform emerged in the very same years when the reform of clock times took off, often with significant overlap between figures involved in both movements. Matters really gained momentum in the first decade of the twentieth century, when the burgeoning movement of international chambers of commerce became the leading force of calendar activism. Calendar reform had started out as an effort to spread the Gregorian calendar but soon turned into a program for introducing a universal "world calendar" across the globe.[33] Calendar reform was more actively supported by businessmen and what could broadly be characterized as capitalist interests than the adoption of mean times had ever been. It was calendar time, not the introduction of mean times, that pointed to a relationship between "abstract" time and capitalism, although not as envisioned by E. P. Thompson and Moishe Postone. In the 1920s and 1930s, now supported by the League of Nations and chambers of commerce alike, the movement reached its zenith.

• • • •

THE CHAPTERS that make up the history presented here take time reform to a variety of sites around the globe. The story begins with the introduction of nationwide mean times in Germany and France. It then proceeds to Britain, where the subsequent step in the debates about uniform clock times played out with particular verve: the discussion about summer or daylight saving time. Next, the story documents the spread of countrywide mean times beyond Europe and North America—in the British Empire, German colonies, and Latin America. The history of colonial and anticolonial time is then explored in more depth through the example of British India. In another non-Western yet non-colonial setting that follows, in the late Ottoman provincial capital of Beirut, Arab intellectuals and reformers were debating the efficient management and use of time. Time reform is then taken to Muslim scholars in the Eastern Mediterranean. The story concludes with an exploration of calendar reform, a diverse movement that encompassed organized business interests, the League of Nations, and many individuals promoting a variety of reformed calendars around the world. The thematic division of labor among these chapters proceeds as a classification of time reform into the different intellectual debates and institutional processes that "time" touched upon around 1900: national, political, and legal time; social time; capitalist time; colonial time; time management; religious time; calendar time.

Language is important, and "naming" time is a feature of the story told here. This book frequently contrasts the adoption of countrywide mean times (most commonly one single time deployed as the mean time throughout a state) with local or, alternatively, solar times. In the period covered, these local and solar times were exclusively *mean* local and solar times—that is, times adjusted for the variations of apparent time (the time of a sundial) over the course of the year. For reasons of clarity and legibility, "solar time" or "local time" are used interchangeably instead of the astronomically accurate "solar mean time" and "local mean time."[34]

"Greenwich time" and the formally correct "Greenwich Mean Time" seldom found application outside of Britain. When referring to nothing more than a numerical time difference between the zero meridian at Greenwich and a mean time or solar time somewhere in the world, this book therefore uses what is today's official designation for time zones, "Coordinated Universal Time" (UTC). That designation is no more "neutral" than is "Common Era" compared to "Christian Era," but at least UTC evokes less specific historical contexts than its alternatives. When talking about the specific and exact times adopted in colonial contexts in particular, the language of the sources is often confusing, as it reflects the origin of particular archival material that happens to report on a certain instance of time change. This is often not the language of colonial administrators in and of that colony itself but the language of those who coincidentally write and engage with this particular colonial setting yet come from an altogether different background themselves.

Modern-day historians writing in English mostly use "standardization" when telling the story of adopting time zones. "Standardization," however, is by no means merely descriptive either but rather a historical term that predominated in the United States and rarely beyond. The term only became widespread with the success of Fordism in the 1920s and 1930s. It is therefore tied to a specific notion of uniformity and uniform time and is here mostly reserved for this particular context. Nineteenth-century observers used terms such as "uniform" and "universal" time and spoke of the "unification" or "reform" of time and calendars. "Uniform" and "universal" carry their own historical baggage, and "reform" is not a neutral term, either. It oozes nineteenth-century notions of the malleability of social behavior and the desire to "correct" humans. Victorian reformers in particular, and their contemporaries in other European societies as well, were fond of such notions when targeting crime, sexuality, and other vices. Yet the language of universalism, uniformity, and reform was the idiom spoken by the countless individuals around the world who sustained the global transformation of time, historical actors without whom the story this book tells is unimaginable.

CHAPTER ONE

# National Times in a Globalizing World

In the second half of the nineteenth century, the world was rapidly becoming more interconnected. People, goods, capital, and ideas moved with growing ease and speed across countries, regions, and continents. In the very same decades, nationalism surged in Europe, bureaucracy and administrations grew in many states, and governments erected legal and economic barriers against the movement of people and goods. These tendencies were similarly central to the process of globalization, as were movements and flows. Time, too, was simultaneously becoming a more global preoccupation, in that certain concepts and ideas spread and moved. At the same time, such concepts and ideas constituted an element of national and regional identity, when those same circulating ideas were anchored in specific contexts to serve national goals.

The interplay of national and global time was a matter of legislative and bureaucratic time. Scientists and other experts drafted and circulated ideas and proposals about global time reform, envisioning an ambitious worldwide system that would supersede particular local and national times. Such information was subsequently picked up in different countries by self-stylized reformers and journalists, who then published newspaper articles and pamphlets on the reform of time from the perspective of Germany, France, or another European country. These publications were often what informed officials in different nation-states on the latest proposals and developments in timekeeping. Yet these lawmakers, bureaucrats, and railway administrators understood uniform time to be a technocratic device in the service of national interests and national administrative rationality. A comparison of French and German politics of time unification illustrates this interrelatedness of globalization, state- and nation-building, and nationalism.[1] Scientists convening at international conferences devised schemes for global time governance, but their ideas required mediation and translation to attain meaning in national politics. Government officials and administrators understood that the world was becoming more interconnected. But to them, the conclusion to draw in the face of globalization was to make the flow of goods, people, and ideas serve the interests of the nation-state. Nationwide mean times were positioning devices that situated a country in its immediate regional and possibly global geopolitical context.

Besides the politics and bureaucracy of time, the French and German stories also engender the social and cultural tenor of time reform, a logic that was always part of nineteenth-century time reorganization: in France, the domestic watch industry managed to turn the preoccupation with accurate and precise uniform time into a business that thrived on ideologies of time thrift and punctuality; in Germany it was less often precision that came up in debates about time and more regularly the social impact of time reform as potentially changing people's behavior. Rationalizing interests of administrators thus overlapped with cultural and social motives for time reform; in many cases, the two were inseparable.

Initially, German and French time politics did not take much note of the other side's doings. That changed in the years immediately preceding and following World War I. Growing antagonism among European powers and deteriorating relations between Germany and France in particular did not fail to leave their mark on time politics. Internationalist meetings and conferences now became a battleground for Franco-German competition. Yet nationalism was not the only factor that shaped the reform of time beyond the level of international conferences and global plans. While nationalism was central to the global spread of mean times, local politics, too, shaped the reorganization of time. The national or nationalist meanings and functions of universal time as identified by government officials and bureaucrats were often met by local demands for unified mean times. Together, this push-and-pull on various levels made for an uneven and multitiered spread of uniform time. Such nested, multidirectional moves characterized not only the process of time unification, but also state-building more broadly.

• • • •

THE NINETEENTH-CENTURY globalizing world was, above all, a world on the move. Migrants relocated on an unprecedented scale from Europe to the New World, soon to be joined by equally impressive numbers of Chinese shipping off to America, Australia, and Southeast Asia. Indians followed the trajectories of labor within the British Empire and worked on plantations in the Caribbean and railway constructions in Africa.[2] Some returned, telling of strange habits and cultures, and often discrimination and gruesome working conditions; others had never intended to see their homes again in the first place and became visible representatives of foreign societies in their adopted countries. Economically, the decades prior to the outbreak of World War I witnessed exceptional economic growth in exports of manufactures and capital, and in certain areas, the convergence of commodity prices and wages.[3]

Imperialism and colonialism constituted another form of interconnectedness and movement. Between 1876 and 1913, roughly a quarter of the globe came under some form of Western colonial power. Imperial imagery and knowledge found their way into popular culture back home, where colonial motives populated advertising or the penny press.[4] Abroad, contacts between colonizing powers and non-Western societies were often replete with violence and coercion, but that does not render them less central to globalization.[5] Nineteenth-century globalization was dominated by the North Atlantic world and Britain in particular. Europe and North America were paramount in politics, in economic development, and in their own imagination at least, in social and cultural trends in much of the world. By 1913, the leading industrial powers of Britain, the United States, Germany, France, Russia, and Italy produced over three-quarters of world manufactures. Many parts of the non-Western world, colonized or pressured by informal imperial expansion, were relegated to supplying raw materials and cheap labor. Britain, the largest colonial power of all, occupied a position central to many dimensions of globalization.[6]

The intensification of migration, economic integration, and political contacts led contemporaries to establish several international organizations, agreements, and institutes that would govern this globalized world. Internationalism, in its several hues uniting socialists, women, eugenicists, and statisticians, to name but a few, aspired to mirror this interconnected world by operating beyond the level of the nation-state.[7] As one such international agreement, universal and uniform time, hailed as a lubricant for a highly interconnected world, was to permit the seamless flow of people, goods, and ideas. Like uniform weights and measures based on the decimal system and standardized rates for mailing letters and sending telegrams, uniform time would establish commensurability and comparability and allow for commodification and exchange.

• • • •

NINETEENTH-CENTURY interconnectedness was underpinned by a newformed network of railways, telegraphs, and steamers, girding the globe ever so tightly in the latter decades of the nineteenth century. Interactions between world regions were by no means new, for different innovations in transportation had previously fueled long-distance trade. In the early modern era, commercial and political exchanges over long distances led to the formation of certain regions such as the Atlantic and Indian Ocean worlds. The slave trade, Manila galleons shuttling between the Philippines and Mexico, and the Silk Road are some examples among many of earlier interconnectedness.

The latter half of the nineteenth century differed from earlier periods and phases of globalization in certain regards. First, the intensification of imperial and colonial competition, most succinctly captured in the term coined to describe the partition of the African continent in this period (the famous "scramble" as contemporaries referred to it), lifted this dimension of interconnectedness to unprecedented levels. Then, in the second half of the nineteenth century it had become nearly impossible to remain "unconnected." In one way or another certain elements in many societies were drawn into the newly forming networks of interactions, so much so that autonomy was unsustainable. Lastly, and importantly in the context of time reform, nineteenth-century contemporaries were keenly aware of just how globalized the world they inhabited had become. Elites in places as different as Europe and North America, Argentina, China, Japan, British India, and the late Ottoman Empire looked around and reflected on a world that was seemingly becoming smaller and, in the eyes of non-Westerners, more and more dominated by European science, ideas, and all too often, brute power. Nineteenth-century globalization was self-reflective.

It was this self-conscious contemplation on the global condition that informed many calls for more uniform methods of timekeeping. Assessments such as "All men are, directly or indirectly, in relation with each other" became a common staple of internationalist talk.[8] Sandford Fleming, a Scottish-Canadian railway engineer who was one of the first to propagate a system of worldwide time zones, spoke of the "twin agencies steam and electricity" that annihilated distances and made reform necessary.[9] Scientists and other advocates of uniform time adopted a language of universalism in shaping international scientific exchange in the latter half of the nineteenth century.[10]

In one installment of this narrative, the administrators of a German literary and cultural society described their motives behind advocating time reform in the following terms: "The greatest successes our age can take pride in are those pertaining to the overcoming of spatial separation—be it that with the help of the telescope it is possible to penetrate the most remote depths of space, be it that on earth, a thought is transmitted through the wire, or that people and goods move on the wings of steam from place to place. Mankind's striving to impart its intellectual gifts and goods, and to exchange the riches of different sites and countries of the earth has increased the need for such an overcoming in ever-higher degrees." Such striving, the authors held, "does not tolerate resistance," and if necessary, "it fights against those peoples who do not partake, who become guilty of the crime of opposing it." Colonial conquest and informal imperial control were inevitable responses to obstructing the progress of globalization. The authors observed that "the people of the West in the old world and their offspring in the new are demanding access to the lush wilderness of Madagascar and to the often strange as much as admirable state of Japan, in the name of forced treaties or persistent incompatibilities." Such justifications of forced "globalization" in the name of progress prefigured what President Theodore Roosevelt would term the "attendant cruelties" of civilization in his vindication of the American military intervention in the Philippines in 1899. It was rare for universalist claims to sit so comfortably and explicitly next to threats, but universalism was never neutral, and talk about the connectivity of the world went beyond impartial analysis. Globalization was not only self-reflective but also an ideology of sorts.

Our authors at the German literary association concluded that interconnectedness, forced or voluntary, mandated more uniform time: "The more spatial separation is overcome, the more general and increased the exchange in intellectual and material messages among all peoples and countries of the earth, the more urgent and important is the need for a general, matching calculation of time, through the precise correctness of its basics permitting certain calculation and determination in any place."[11] Such calls, whether for uniform clock time or, as in this case, calendar time, were common among legislators, scientists, self-styled calendar and clock time reformers, and even watchmakers. The plurality of ways of counting, measuring, and valuating clock, calendar, and social time was by no means new. While most societies appear to have had a desire to mark time, they did so in myriad contrasting ways throughout history and scattered over the world. In an interconnected world where a growing number of people wound up on the move in one form or another—as part of imperial and colonial endeavors, as entrepreneurs, as migrants, missionaries, or pilgrims, for instance—such an existing heterogeneity of times now was more visible than ever before.

Prior to the advent of standard time, church towers, town halls, and train stations in Europe and North America kept solar time. Noon was marked when the sun crossed the meridian at a given location. Every city, town, and village theoretically observed a different time depending on its longitudinal position. Solar time—"true time," as Americans referred to it—was often not much more than an educated approximation. Determining accurate local time and, later, mean time required expensive precision tools such as accurate clocks, astronomical instruments, and observatory charts.[12] Yet even where it was possible to determine solar time with some degree of certitude, the sun itself was and is an imperfect timekeeper. The earth's axis is slightly tilted, and its orbit is not circular. The sun therefore moves closer to the earth in the northern winter than in the summer. As a consequence, solar days vary in length depending on the season. On the other hand, a mechanical clock advances uniformly. With the spread of clocks and more slowly, watches, the local time on display in every village or town became mean solar time in the early nineteenth century, that is, a time that corrects for the variations of apparent solar time as measured by a sundial.

For many decades, the coexistence of multiple solar times affected only a small number of people who traversed long distances. When around the middle of the nineteenth century railways made travel easier, faster, and more affordable, this changed. In certain connection hubs, travelers now had to calculate their way through a thicket of times kept on different lines. In 1875, about seventy-five railway times were used in the United States, six alone in Saint Louis, five in Kansas City, and three in Chicago.[13] Railway men especially in the United States and, to a lesser degree, in Europe were savvy and connected enough to join forces with observatories and survey departments to promote the cause of uniform time zones. Internationalist activism for time standardization proliferated in the second half of the nineteenth century. Professional associations, national scientific academies, and unaffiliated individuals produced a mounting number of pamphlets and articles on the scientific aspects of timekeeping and time unification. Specialists on the topic read each other's publications, often excerpted and referenced internationally in the swelling volumes of scientific and popular-scientific journals and proceedings published in Europe and North America.

Several conferences and meetings of scientific associations promised to make the adoption of uniform time more than just a paper tiger. In 1883, astronomers, geodesists, and surveyors at the International Geodetic Association's conference in Rome discussed the prospect of adopting a prime meridian on which to base a system of uniform hour-wide time zones. In the late nineteenth century, maps and ephemerides of different national origin were based on a variety of meridians. While many survey departments and mapmakers used the meridian of Britain's observatory at Greenwich, the meridians of Paris, Ferro, and Cadiz represented only a small selection of others in use.[14] Jerusalem was actively promoted especially by scientists and others of Catholic background; during the phase of frequent scientific congresses in the early 1880s, a French citizen wrote to the German emperor and suggested that Wilhelm II should back Bethlehem to serve as prime meridian, a deed "worthy of Charlemagne."[15] Conferences like the one in Rome addressed this plurality of meridians and proposed more uniform solutions instead. In closing, the Rome conference commended governments to adopt the "Greenwich Meridian as the initial meridian, . . . as, from a scientific point of view, it fulfills all the desired requirements, and currently being the most widespread of all, carries most chances for general acceptance."[16] It was true that many maps and charts used Greenwich; but Britain's national observatory, perched on a hilltop in the lush greens of southeast London, did not equally appeal to everybody. The island of Ferro, the smallest of the Canary Islands, lured others who insisted that a universal meridian ought not run through territory and should possess no national qualities at all.

Another important meeting was not an international but a national American affair. In the spring of 1883, North American railroads switched to Standard Railway Time, consisting of four hour-wide time zones stretching across the American continent. Implicitly, these zones were counted from a zero meridian running through the observatory at Greenwich, United Kingdom.[17] In the aftermath, scientists and railway men knew they had to enlist the support of national governments if their plans were to have clout. In the early 1880s, a group of such figures led by the American Meteorological Society did just that in lobbying US government agencies to prepare an international conference devoted to uniform time. Two years later, President Chester A. Arthur welcomed delegates from twenty-six countries to Washington, DC (invitations had been extended to all countries that maintained diplomatic relations with the United States at the time) for the Prime Meridian Conference of 1884. Among the attendants figured a few diplomats, but most participants were members of national observatories and thus, broadly speaking, scientists. After roughly a month of negotiations, the conference voted to adopt Greenwich as the prime meridian of a system of twenty-four hour-wide time zones. Of the countries in attendance, only Brazil and Haiti voted against Greenwich. France abstained.[18]

• • • •

INTERNATIONAL EFFORTS to coordinate time certainly mattered in the eyes of those scientists who partook in the flurry of conferences and meetings in these years. But among the ranks of different bureaucracies in various European states, the impact of international deliberations was barely palpable. Outside scientific circles, the unification of time and the introduction of time zones were understood primarily as matters of regional integration and national state-building. West and northern European nation-states adopting single zone times in the late nineteenth century mostly did so with their minds set on improving train traffic and streamlining administrative processes within their states. Scientific and technological arguments for more accurate and reliable time resonated with officials who found themselves attracted to fantasies of accuracy and precision, a "trust in numbers," as one historian puts it. Uniform time appealed to the cast of technocratic-minded ministers and bureaucrats who staffed many public works, postal, telegraph, army, and education departments from the middle of the nineteenth century. In their view, accurate time was just as important for social engineering as accurate data was for purposes of social management and the prevention of crime.[19] But these tools were to be harnessed for the engineering and improvement of national societies, not the facilitation of global connections.

The regionalization and nationalization of time and even its infusion with geopolitical significance can be observed in another important feature of nineteenth-century time reform. The term "universal time" barely found application outside of scientific-internationalist conferences and pamphlets. Occasionally, German participants in internationalist discussions deployed what could perhaps be understood as the German-language equivalent of universal time, so-called World Time (Weltzeit). Even Greenwich time, common in British discussions but despite the system's global prevalence in the very long run, was hardly used in countries other than Britain. French legislators preferred to speak of "legal time" and "official time" when referring to any nationwide mean time. Other Romance languages such as Italian, Portuguese, and Spanish followed the French pattern in mostly using "legal time" and "official time." In Germany, where idiosyncratic reactions to Greenwich did not exist, with a few punctual exceptions, nobody spoke of "Greenwich time." In the United States, it was not Greenwich time but rather standard time that was discussed, eventually becoming a predominantly American expression.

Most indicative of the regionalization and nationalization of universal time was arguably the tendency to designate time zones as mapping geographical notions. Austrian plans to unify time first on railways and then in civil life were accompanied by calls to name the time one hour ahead of Greenwich "Adria-Zeit," or Adriatic Time. The new zone would cover countries such as Austria-Hungary, Germany, Serbia, Switzerland, and Italy, in the views of its proponents thereby conveniently branding the imagined geopolitical space of southeastern Europe. Other proposals such as "Balkan Time" were dismissed but suggested a similar affinity for concrete geographies in lieu of abstract universalities. Robert Schram, author of several articles on Adriatic Time, moreover proposed that the time two hours ahead of Greenwich be called "Bosporus Time," as it would extend to Russia, Romania, Bulgaria, parts of the Ottoman Empire, and Egypt. In France, the time zone two hours ahead of Greenwich was frequently called "Oriental Time." American time zones too had obtained such geographical denominations, although less openly politicized ones.[20]

Germans rendered time into political space in the most telltale manner. When German officials discussed the adoption of a new nationwide mean time, they most frequently termed it "Mitteleuropäische Zeit." The expression translates poorly into English as "Central European Time," the most commonly found English-language rendering today and in the nineteenth century. "Mitteleuropa" (Central Europe) was not simply a geographic denomination as the English translation might suggest. The interconnected world of the nineteenth century not only generated a reconceptualization of global time but also a new way of thinking about global space. Geopolitics, as pioneered by the British Halford Mackinder and the American Alfred Mahan, had a number of influential acolytes in Germany. Here, it was the writings of Friedrich Ratzel (1844–1904) and later Albrecht Haushofer (1903–1945) that shaped ideas about the twin nature of politics and geography and the formative forces of geography as shaping history and destiny. "Central Europe" or "Mitteleuropa" was a key term coined by contemporary writers to denote Germany's and German-speaking Europe's "middling position" (Mittellage). Mitteleuropa stretched from the North and Baltic Seas to the Adriatic Sea and the southern fringes of the Danube plain; from the river Vistula flowing through present-day Poland in the east and the Vosges mountains in France in the west; from the Baltic to the Balkans.[21]

Even in the late nineteenth century, when Germany was much bigger than after 1945, some of these areas lay outside the borders of imperial Germany. Central Europe and Central European Time, then, were geopolitical concepts about the position of Germany within Europe and its political, economic, and even racial extension into the east.[22] The new time might have even suggested that Germany's "late" national unification, which came only in 1871, had now been overcome by uniting the formerly fractured space of Central Europe under a new time that signaled political and bureaucratic modernity. That Central Europe was not merely a descriptive but highly normative notion became apparent in 1915. After the outbreak of the war, the German publicist Friedrich Naumann published an eponymous book, *Mitteleuropa,* that called Germany and Austria-Hungary to the task of forming a veritable Central European geopolitical unit against both Russia and the "Western" alliance of France and Britain. A little more than twenty years later, the far-right jurist and political theorist Carl Schmitt expanded on these geopolitical views. According to Schmitt, Nazi Germany, through its annexations in Austria and Czechoslovakia, was in the process of establishing a juridico-political "large space" (Grossraum) in Central Europe that would guarantee the protection of ethnic German minorities from foreign interference. It was merely another step from such conceptions to Nazi "Lebensraum" in the East.[23]

In conceptualizing geopolitical units such as Mitteleuropa, nineteenth-century thinkers pondered precisely the relationship between interconnectedness and the formation of regions that similarly characterized the reform of time around 1900. Exchange and interactions on a worldwide scale, "Weltverkehr" or "Worldwide Interchange," shaped geopolitics. As Naumann put it, it was "the age of exchange and interaction on a world scale that produced World States." He continued, "Such new technologies like steam and electricity do not work with state entities that came into being under the influence of previous forms of worldwide interchange."[24] In Naumann's version of the nineteenth-century global village, interconnectedness required and simultaneously produced a number of dominant large powers and larger geopolitical units, one of which was Mitteleuropa. German administrators, scientists, and journalists habitually used the term "mitteleuropäisch" when referring to Germany's time zone, without seemingly giving it much thought or discussion. Some of the earliest uses of "Central European" as a geographical/geopolitical designation enveloping the two countries occurred in the late 1880s and 1890s in debates about uniform time, although the concept really only gained broader recognition in the first years of the twentieth century and following the publication of Naumann's book. It might indeed be that the construction of a Central European time zone was what really turned the term into a household name that was then popularized by political geographers in the following decades.

Besides such national and regional designations of time, mean times were contrived to fulfill local functions. Efforts to streamline bureaucracy and improve populations within political and geopolitical units seldom emanated from the top alone. When it came to implementing uniform time within single states, the variety of actors who simultaneously took an interest in abolishing local times is striking. From the 1870s through the 1890s, national or federal governments were not the only ones preoccupied with creating landscapes of uniform clock time. There were municipal authorities, for instance; in cities such as Paris, Vienna, and Berlin, officials began to ponder solutions for synchronizing the numerous public clocks on display throughout these metropolises. Most solutions to the problem of synchronization consisted of a so-called master clock, corrected via electricity by a nearby observatory, and numerous connected clocks often referred to as "slave clocks." One promoter of a system of city clocks for Vienna described such installations as a method by which "as many clocks as possible cease being independent clocks when they give up their individuality and become part of a whole, a big 'clock state' (Uhrenstaat) with a head of state at the top."[25]

The local synchronization of time in cities provided important impulses for the unification of time on the national level. Paris was among the earliest cities to install a synchronization system in the 1870s. It opted for a complicated pneumatic mechanism that used a master clock but then sent not electric pulses but steam pressure through a network of underground pipes. As any such system, pneumatic time synchronization was prone to malfunction. Parisian archives tell a story marred by dysfunction arising from disconnected clocks and a host of other problems.[26] Faced with such challenges, local bodies were therefore at times more insistent than national governments in procuring uniform *national* time for their cities. When different railways in Germany and Austria-Hungary began to apply the time one hour faster than Greenwich on their lines, cities and towns in the surrounding areas singlehandedly adopted the new time as well. A handful of Austrian, German, and smaller regional French cities thus obtained what would become the nationwide time standard prior to the unification of national time.

In other cases, European states would introduce nationwide mean times in the 1880s and 1890s by initially choosing the time of the capital, thus disregarding the stipulations of international time reformers. Then a decade or more later, once a neighboring country had perhaps adopted a nationwide mean time in sync with the grid of hour-wide zones, that time would be changed to a mean time differing from the meridian of Greenwich by even hours rather than minutes and seconds. The fact that in the closing decades of the nineteenth century, a handful of northern, central, and western European countries (as well as Japan) did choose mean times in accordance with the system of time zones should therefore not be viewed as a foregone conclusion. Sweden, always adduced by German authorities as an example of successful, early change, introduced the time one hour ahead of Greenwich independent of international deliberations in 1879; Japan, particularly eager for reform at the beginning of the Meiji Restoration, used UTC+9 as of 1888. Denmark, Italy, Norway, and Switzerland followed suit in the mid-1890s.[27] Such outcomes were hardly preordained. The emergence of nationwide mean times occurred as a cumulative, gradual, at times highly contingent process often characterized by the absence of directionality. The adoption of nationwide mean times only stood at the end of a centrifugal dynamic in which municipal and regional initiatives were eventually capped off by national policies designed to put a stop to unilateral local and regional action. State-building appears here as a much more lateral, multileveled process than a top-down one. It has been argued that nation-states, even allegedly centralized France, in fact appear to have been "largely built from the margins and from the bottom up." The process of adopting countrywide mean times seems to corroborate such findings.[28]

• • • •

THE GOALS of the railway administrators and local and national government officials who carried out these time changes seldom chimed with the ambitions of international scientists. Yet it is nevertheless possible to discern a connection between international conferences and the circulation of knowledge, on the one hand, and the unification of local and national times in various countries, on the other. Congresses, activist pamphlets, and other publicity brought time unification to the attention of government authorities. Robert Schram's newspaper essays and pamphlets are a case in point: in the early 1890s, Schram, the aforementioned Austrian geodesist based in Vienna, was following the exchanges about improvements in timekeeping in the popular scientific press. At home in Austria, railway officials and bureaucrats at this point were still discussing whether to make Vienna time the country's official legal time, taking no notice of worldwide time zones. Schram placed a series of articles in the daily press and mailed off numerous pamphlets to government authorities, and instead advanced a plan to choose the time one hour ahead of Greenwich as an implicit extension of the American-favored system that had been introduced in the United States in 1883. When Austrian railway officials announced their decision to opt for a time zone based on Greenwich instead of Vienna time, it was Schram's work and proposals that were quoted approvingly in the explanations.[29]

Schram and countless other reformers tirelessly penned articles in the daily press as well as shorter pamphlets on the causes they advanced. Together with offprints from journal articles, these little brochures were mailed by their authors unsolicited to government ministries and other official authorities. Schram's own correspondence, for instance, surfaced in German and French government archives on time unification as well as in the archives of the Greenwich Observatory. The relatively cheap dissemination of self-published pamphlets became one of the most important vehicles of mediating between different spheres of governments, scientists, publicists, and reformers and of spreading uniform time—or whatever "uniform" meant in single national and regional political and cultural contexts.

Schram may not have been the only voice to verbalize such ideas; other factors possibly contributed to the Austrian railway office's decision as well.

But scientists and observers such as Robert Schram serve as an example for the way in which ideas moved and were subsequently appropriated for different purposes, with different functions in different contexts. International epistemic communities like that of the time unifiers facilitated the exchange of ideas and concepts via improved, accelerated, and cheapened means of (print) communication and transportation, and individuals like Schram played a crucial role in mediating between the worlds of science, railways, and government, in translating the concerns of one side into the interests of another. The movement of ideas, concepts, and practices in nineteenth-century globalization seldom percolated from global to local contexts on a horizontal trajectory. When ideas about time moved, the channels of transmission for concepts and notions of time mostly moved in tracks that did not map onto scalar hierarchies reaching from the level of global reform schemes and international conferences to regions, nation-states, and local contexts. Instead, in an interconnected world, ideas and practices tended to occupy a middle ground and moved laterally rather than vertically, from intellectuals and thinkers in one society via print products such as pamphlets and newspapers to their counterparts in another. In many contexts, intellectuals, journalists, and publicists functioned as translators between the sphere of international conference deliberations and local societies. A globalized world offered Schram and others the means for regionalizing and nationalizing time.

• • • •

When Schram nudged Austro-Hungarian railways and government officials to adopt Central European Time, he inadvertently set off German time unification as well. Germany only turned into a nation-state of sorts in the second half of the nineteenth century when, under Prussian and notably Otto von Bismarck's leadership, Germany's numerous duchies, grand duchies, kingdoms, principalities, and other territorial states and cities were, often forcefully, incorporated into what was now the German Empire, a process that came to an end with the German victory over France in 1871 and the proclamation of a German emperor in the Hall of Mirrors in the Chateau of Versailles. In a now-unified Germany, however, federal and regional structures remained powerful both as a source of identity and the seat of politics. Time was one area where regionalism and unorthodox heterogeneity prevailed.

In the winter of 1889, German officials launched a major inquiry into the question of adopting a standard time on all railways. While the topic was not new, it had gained traction from a petition by the Association of Hungarian Railways to use the time of the fifteenth degree east of Greenwich (one hour fast) on railways within its purview. In order to render the Austro-Hungarian plan more efficient, the president of the Hungarian state railways approached German officials in the hope of convincing neighboring countries to pursue a similar time change. The times currently operated on German railways mirrored the strong regionalist traditions of the country. Most northern German railways as well as those in recently annexed Alsace Lorraine followed Berlin time or, as one might say, Prussian time. In the southern half of Germany, trains ran on other regional times: Munich, Stuttgart, Karlsruhe, Ludwigshafen, and Frankfurt time, respectively. Where lines pertaining to different systems met, travelers had to convert the times.[30] Railways continued to be organized on the level of the German states rather than nationality, even after 1871; only in 1920 were railways fully nationalized. Since the organization of railways was in the hands of the states, differences in pricing, scheduling, and ownership were unavoidable. Many individual German states did nationalize railways in this period; still, several private lines remained. Often, the importance and benefits accorded by railways were therefore reaped in individual states and much less so on the national level—another indicator of the nested and uneven nature of state- and nation-building.[31]

In light of the plurality of German times, the Hungarian proposal was viewed with interest but also considerable skepticism. It was poised to do away with some of the cumbersome calculations required for switching between different lines in certain connection hubs. But it was bound to create new problems as well. If standard time were to be extended into everyday life, the difference between local sun time and the new mean time would cause confusion and disarray. In the late nineteenth century, deviations of mean time from solar time by a mere fifteen minutes were considered a major obstacle to time unification. When discussing the Hungarian plan, critics therefore pointed out that roughly a quarter of the German population would have to "endure" a difference of more than twenty minutes between solar time and mean time if Central European Time were used in daily life. After more discussions, the general assembly of the Association of German Railways agreed to operate Central European Time only internally and for administrative purposes on all lines within the association's purview as of 1891.[32]

In this situation, Prussian officials began to informally ask whether an extension beyond railways would be beneficial and whether, generally speaking, such a thing as "time" as an aspect of everyday life could be the subject of legislation at all.[33] Passing a law with the goal to affect people's everyday schedules and sociotemporal habits was never a logical step but rather subject to much discussion, in Germany and beyond. Many contemporaries were firmly convinced that time and especially individual temporal behavior could and should not be regulated by law. After initial debates, railroaders stopped short of actively seeking to extend the use of Central European Time externally to civilian life, but recommended the move for the near future. There was a lack of unanimity over how to proceed from there. Some advocated letting the public decide whether to follow the new time; others proposed passing legislation in bigger cities but leaving it up to smaller towns and villages to adjust their statutes and practices.

In these talks, making Central European Time the legal time in all of Germany was not even the default solution. Observers found no problem with envisioning the parallel application of multiple times, one administrative and official, the other more closely tied to everyday life. The practice of keeping parallel times was in fact already common practice. In several towns, clocks in courthouses were set back ten minutes to grant those summoned a grace period for accommodating possible delays. Train stations set clocks back by as much as fifteen minutes, and local government offices often did so too, thus creating a multitude of informal timing practices.[34] German officials even toyed with the idea of equipping station clocks with two hands, one for train time or mean time, one for local time.[35]

Remaining differences in opinion notwithstanding, officials eventually acceded to placing a proposal for uniform "zone time" (Zonenzeit) before the German parliament. When almost another year later the Reichstag discussed the railway department's views on uniform times, it was Helmuth von Moltke's time to shine. As a result of the shrewd strategist's impassioned speech, German "Einheitszeit" or uniform time was now so closely associated with Moltke that contemporaries, perhaps not entirely tongue-in-cheek, demanded that the new time be officially designated "Moltke-Einheits-Zeit," or Moltke Uniform Time.[36] Making the connection between railways and the German nation-state as Moltke did was not the general's invention by all means. It had been the German economist Friedrich List who, in the 1840s, famously encouraged the construction of railways as an "ingredient for the fortification of the national spirit."[37] After 1871, German officials indeed hoped to place railways under federal control and deploy them in the service of state- and nation-building, only to see their aspirations dashed by the assertion of state rights and the persistence of federalism in the organization of railways.[38] Coming from Moltke's mouth half a century later, invocations of railway time as a purveyor and emblem of national identity became an incontrovertible truth.

Moltke's presentation before German legislators in April 1891 first convinced Prussian authorities. As of June 1891, Prussian railway officials threw out Berlin time and replaced it with Central European Time for internal, administrative purposes. To the public and to postal services and the military, authorities still eagerly dispensed brochures and timetables listing local times. In Austria-Hungary, where railways had dropped Prague and Budapest times and adopted Central European Time both internally and externally, several communities and municipalities surrounding major railway lines, among them Salzburg, Graz, and Trieste, unilaterally and independently of empire-wide legislation switched to the new time.[39]

Time unification in Germany continued with the adoption of Central European Time on southern German railways in 1892. When northern German states declared to do the same as of April 1893, the imperial German government reacted by passing a law that elevated the time one hour ahead of Greenwich to the status of Germany's official time for all purposes. On paper, Germany now followed a single uniform time.[40] In preparation for April 1, the railway ministry instructed station personnel on the necessary change of times. Any such practices as delaying station and courthouse clocks to assist a tardy public, the Ministry of Railways explained, were rendered illegal by the new law and had to cease.[41]

Upon weighing uniform time for Germany, officials in all branches of regional and imperial administrations turned to other countries where time had previously been unified in one way or another. The most frequent reference was made to Sweden, where the time one hour ahead of Greenwich had been adopted as early as 1879. Neighboring France, where Paris time had been introduced on all railways as of 1889, also featured in German discussions. It did not matter that the French choice was supposedly in violation of the system chosen by a large group of observatory officials, scientists, and diplomats in the American capital a few years earlier. Nobody even appeared to take note of the breach. And not once did it occur to German government officials that the Washington Prime Meridian Conference in 1884 had laid down rules and established a precedent now to be honored. The conference was not even mentioned. Time was a national affair. The Washington, DC, gathering of 1884 remained largely without impact.

This was not least the result of attitudes among governments themselves. Already at the preparatory state, the general mistrust of internationalist activities at the level of national governments was apparent. The prospect of introducing a nationwide uniform time that would operate outside of railways and telegraphs, and do so at the behest of an international gathering of scientists and diplomats, was alarming to German administrators. During the conference of 1884, the idea of an "all-encompassing" reform of timekeeping, with nationwide mean time applied to all aspects of daily life, appeared "barely practical" to German officials.[42] Temporarily, voices cautioned against participating in the conference at all, given its unclear and vague competencies. But soon, cynicism prevailed. Germany dispatched two delegates to the 1884 Prime Meridian Conference, both diplomats. When the German Foreign Office briefed its representatives, it instructed diplomats to steer clear from offering support for any decision that would impose the adoption of a certain time for civil life. Any resolve thereon had to originate with the respective domestic authorities and would precede an international mandate. During the conference, German officials in Berlin repeatedly assured one another that a resolution, if adopted, would not be binding anyway and therefore did not really matter.[43]

In the discussions about national time reform in Germany, one caveat transpired that would take on a dynamic of its own roughly thirty years later. The relationship of mean time to solar time and thus the relationship of standard time to daylight was a much contemplated factor already in the adoption of nationwide mean times and time zones throughout the 1890s. Tied to this was a genuine inability to imagine time and time zones as abstract conventions, concepts grafted onto the natural rhythms of day and night, daylight and darkness. Contemporaries experienced great difficulties imagining a world where the rhythms of life were no longer determined by the sun. This led many administrators and observers involved with time unification to ponder measures for coping with the switch to mean time by making it like sun time. Some commentators were convinced that such steps were the natural conclusion to any course of action that created divergence between solar time and mean time. Officials, for instance, debated whether work laws would have to be amended to reflect the switch to nationwide mean time and to retain the relationship between working hours and daylight in different regions, even if the switch entailed nothing more than a fifteen- or twenty-minute change.

Long before daylight saving became an issue officially, the abolishment of solar time was coming to be viewed as the social question it would be in the opening years of the twentieth century. Prior to the adoption of a nationwide mean time, unofficial daylight saving schemes were operated in many schools and factories in Germany. During the summer, starting hours were made to fall earlier in order to guarantee the most beneficial use of daylight.[44] In Belgium, opponents of mean time calculated the amount of petroleum likely to be burned as a consequence of more adverse lighting conditions if clocks were set back seventeen minutes—the difference between Brussels time, the country's previous nationwide mean time, and the new standard time.[45] In the context of introducing the time one hour in advance of Greenwich, officials in the Netherlands contemplated altering the business hours for schools, government offices, and railways in legal statutes, in order to retain activities at roughly the same solar time as previously.[46] Another open question to Dutch administrators was which time zone to eventually join, Germany's Central European Time or Britain and France's time of one hour less. The case for German time was made by underscoring the close commercial and economic ties to the eastern neighbor; the case for pivoting westward was made by pointing to the medical and social benefits of gaining daylight through the adoption of a time that was further ahead of solar time and had the effect of advancing the clock akin to future daylight saving measures.[47]

German legislators bemoaned that in some regions workers would have to leave for work at 5:30 instead of 6:00 a.m., thus imposing undue hardship on the laboring population. Instead of taking 5:30 to be the new 6:00 a.m., such arguments seemed to assume that the "real" 5:30 would always continue to be what it was, contrasting abstract and movable clock time with absolute and immutable "real" time.[48] Nineteenth- and early twentieth-century Europeans placed great importance on the interval between the daily routine of rising, having breakfast, and venturing off to work. Nothing irked contemporaries more than when these routines were disrupted by meddling with the time intervals between morning drills and the position of the sun—that is, the time when daylight rose.

When southern German states introduced Central European Time on regional railways for external as well as internal purposes in 1892, imaginary hurdles to time unification suddenly became very practical. Railway companies had in fact proceeded to change their schedules in a way that would have trains run at the same solar time even after mean time had been introduced. Confusion arose as people noted that only the nominal designation of arrival and departure had changed—hence in a location where the new mean time was half an hour fast on solar time, a train that previously departed at 9:00 a.m. was now rescheduled to run at 9:30. But work times often had been shifted to mean time without such adjustments. As a consequence, workers no longer encountered trains timed to take them to work at rush hour in the morning. As an expression of general confusion, people could be heard uttering, "My watch shows 9:30, hence it is now 9:00." Such perplexities would linger for decades.[49]

• • • •

As in Germany, French politics of unifying nationwide time were informed by deeply rooted national historical traditions. In France, it was not the lack of a nation-state but a scientific tradition perceived to be inherently French that was foregrounded. The nation prided itself on a long-standing list of achievements and expertise in time reform, harkening back to the establishment of the Paris Observatory (1667) and the days of the French Revolution alike. During the revolution, a new calendar was instated in 1793 under which the year was divided into twelve months of three ten-day weeks, in order to fully comply with the decimal order brought on by the new metric system. Days consisted of ten hours of one-hundred decimal minutes. Although the reform widely failed to gain acceptance and was abolished in 1806, French commentary on time reform repeatedly evinced the scientific glories of the nation's revolutionary past as a tradition that ought not be discarded lightly.[50] France, contrary to Germany and Britain, was therefore more attentive to international meetings and congresses that threatened to make a meridian other than the national Paris one the starting point of a worldwide system of time zones.

After the American invitation for the Washington Prime Meridian Conference was received in France, a commission was set up at the Bureau of Longitudes and the Paris Observatory, charged with studying the possibility of adopting a time based on Greenwich.[51] To counter the meridian of Greenwich, the commission concluded it best to advocate a neutral, non-territorial prime meridian for the new system and instructed its representatives in Washington accordingly. At the Washington conference in 1884, however, it was soon obvious to Jules Janssen, a member of the Paris Observatory, and to the French consul that insistence on neutrality was bound to fail. "From the first session on," the consul chafed, "there were unambiguous signs that forced me and Mr. Janssen to acknowledge the absolute dominance of the assembly by England and the United States."[52] What may have added to this impression was that a number of countries such as Guatemala and Paraguay, while nominally participating in their own right, were represented by American nationals, a practice frequently encountered in nineteenth-century internationalism.

France's engagement with internationalist time reform first and foremost sparked an interest in national time. During the 1880s and 1890s, when time reform was attracting attention internationally, several times coexisted in France. Three times were officially recognized. In addition to Paris time used on telegraphs and railways, local mean times remained widespread. Like in Germany, railway companies strategically delayed the time displayed on train station buildings, granting passengers a five-minute delay on station clocks inside the buildings in order to compensate for their assumed tardiness. Clocks mounted on the outside of station buildings ran on local time. Many cities added a third clock hand to prominent public clocks to be set to Paris time. Reminiscent of Austria, an observer in the Bordeaux region noticed how all cities along the main railway tracks had informally adopted Paris time, whereas other villages stuck to local times. Thus, within a one-kilometer radius, time could differ by as much as fifteen minutes from one village to another.[53]

In the 1890s, the multitude of times came to be viewed as detrimental to the conscientious and responsible management of time. As in the United States, French watchmakers and observatories capitalized on the newfound commodification of accurate time. Observatories offered various time services against remuneration; clock- and watchmakers cashed in as well. An article on the French watchmaking industry confirmed the value newly bestowed on time. "Never before has the English proverb 'Time is money' been so true," the author declared. "We live fast, we use every minute, and in order to use time rationally, we have to be able to measure it."[54] A train departing at 7:43 does not wait for a passenger who arrives at 7:44, the article warned. And while precision in time was not needed until recently, "today we have, without apprehension, undergone a transformation of our habits." Hence, the conclusion that "time is precious to all of us. . . . Thus, in whatever way we look at time, whether in its proper form or as a monetary quantity, it is an element of considerable value."[55] Observatories such as Besançon, Neuchâtel, and Geneva offered the precision-craving owner of upscale timepieces a regular checkup service. Whoever was willing to separate from a clock or watch for two to six weeks could, for a fee, leave it with professionals for observation. Afterward, the proud owner received a certificate documenting accuracy and functionality of his watch, with the most exact instruments even receiving a mark of distinction at some observatories. Observatories held annual contests in which watches were categorized and prized according to their precision.[56] The nationwide reform of mean times in France was claimed to bear a direct link to such moral refinement; unifying time measurement equaled "truly humanitarian work."[57]

To rid the country of its multiple times and create a uniform French national time in honor of scientific achievements past, a law proposal surfaced. In 1888 it was suggested to make the time of the Paris meridian, famously associated with the astronomic glories of the Paris Observatory, the official time for all French railways and all aspects of public life. But matters stalled before any vote had been passed. Similar to what occurred in Austria-Hungary a few years later, different local bodies and towns now acted independently. The municipal council of Langres in the region of Haute Marne unilaterally adopted Paris time for the city.[58] The Société Scientifique Flammarion in Marseille made a collective move in favor of adopting Paris time in France's provinces and began rallying other local and regional associations.[59] Yet another municipal council, in Marseille, similarly voted to run Paris time on the city's clocks. In the region of Côte d'Or, the head of the employers association addressed the local provincial council with a request to push the French Parliament for a law introducing Paris time as the time for all of France; the application of two different times (regional Dijon time and Paris time) was now said to be the cause of great confusion.[60] Another such demand came from the municipal council of Lyon. Back in Paris, the Bureau of Longitudes began to worry. One of France's state-run scientific institutions active in geodesy and now increasingly timekeeping, the Bureau feared its influence was waning and, in an attempt to maintain an upper hand in all things time, urged the Ministry of Public Instruction to take up the law project anew. A draft bill was submitted to the French Chamber of Deputies in March 1890 and signed into law roughly a year's time later. Paris time rather than Greenwich time became the nation's legal time in March 1891.[61] Internationalism had fostered the creation of national time in France.

Throughout the 1890s, voices calling on France to adopt a mean time in accordance with the time zone system could be heard occasionally. France, together with other western European countries, including the Netherlands, Spain, and Portugal, continued to subsist on non-Greenwich-based zones. In 1897, a draft bill was put before the French Parliament stating that "the official prime meridian used in France is the Greenwich meridian."[62] With hostility toward "British" time still running high, the proposal caused outrage. Deputies together with the press chided the bill for its brazen attempt to "abandon" the honorable Paris meridian and for pushing French commercial cartography (relying on the Paris meridian) to the brink of ruin, all the while "sanctifying the preeminence and superiority of [foreign] scientific instruments."[63] In the same years throughout the late 1890s, the Bureau of Longitudes moreover revived a project for the promotion of "French" scientific achievements that had first been advanced in the months prior to the 1884 conference in Washington, DC, but never came to fruition. Back then a commission had been established that was tasked with the study of decimal time and of plans to abolish the common day of twenty-four hours of sixty minutes' length and instead use hours containing a hundred minutes. Faced with the possibility of eventually following the "British" standard of Greenwich time, French scientists unearthed these old discussions and again debated the decimalization of time units in the closing years of the nineteenth century. When the French foreign ministry approached other European countries with ideas for an international conference on the decimalization of time, responses were couched in the politesse of diplomatic correspondence, but left little room for interpretation: there was no interest whatsoever in any such effort. The French decimalization project was shelved.[64]

Before the French Chamber could fully discuss the proposal to introduce the shameful time based on the Greenwich meridian in 1897, another deputy was quick to advance a counter-project, providing that legal time in France shall be "Paris time, running slow nine minutes and twenty-one seconds."[65] As everybody could all too easily tell, "Paris minus nine minutes and twenty-one seconds" in fact meant Greenwich Mean Time—only that the embarrassing British origin of the system had been obscured.[66] Once the French Senate became involved and likewise had to cast a vote, matters stalled once again. A commission appointed to evaluate the bill wound up taking no less than twelve years to take a stance.

The commission finally published its report on the proposal to introduce "Greenwich time"—albeit without pronouncing the despised British provenience—in November 1910. In February 1911, when the bill reached the Senate floor, reform adversaries attempted to uphold the glory of Paris in the face of British predominance one last time. To prevent the inevitable, Greenwich time opponents had concocted a complicated scheme that returned to the coexistence of multiple times. The plan entailed clinging to the Paris meridian for civilian mean time but in an additional clause stipulated that "Paris time minus nine minutes and twenty-one seconds" would be used on international postal and telegraphic services as well as railroads and internal train station clocks. External clocks were to remain on Paris time. A formula so convoluted in its practical ramifications failed to convince even the most ardent French patriots. On March 9, 1911, the French Senate eventually adopted the law proposal, and the time of the Greenwich meridian was implemented nationwide only two days later.[67]

• • • •

AT THE INITIAL STAGE, German and French politics of nationwide mean time unification proceeded alongside each other without much interaction. French eyes were turned toward the competition with Britain while German thoughts focused on internal temporal heterogeneity and, occasionally, on neighboring Austria-Hungary and central European railways. German and French time politics only crossed paths in the decade prior to World War I. By this point, great power politics in Europe had soured over the festering problem of nationalism in the Hapsburg and Ottoman Empires, colonial claims, the Anglo-German arms race, and the hardening of the European system of alliances, in which France, Britain, and Russia now fell firmly on one side with Germany and Austria-Hungary squarely in the opposite camp.

Internationalism, regardless of its proclaimed attempt to organize interests beyond the petty concerns of individual nation-states, had been a sparring ground for competing national interests from its very beginnings. This tendency was only enhanced by the close interrelation between science and the state during the nineteenth and early twentieth century. The German Technisch-Physikalische Reichsanstalt (the Imperial Technical-Physical Institute), among the world's foremost institutions of physical research at the time, proved to be one such place where science, industry, and the state blended together for the purpose of engineering a modern nation.[68] In France, the geographic division of the army stood for the marriage of science to the state. The French army moreover ran the Ecole Polytechnique, the nation's finest scientific school that perfected the science of land measurement as an indispensable tool for the exploration and conquest of territories.[69] Astronomy had always been closely tied to these endeavors, especially since the establishment of the Bureau of Longitudes in 1795. The Bureau resided at the Paris Observatory and oversaw the totality of France's observatories while publishing a well-respected ephemeris.[70] Nineteenth-century science brimmed with nationalist overtones, and conferences and congresses marked the grounds on which scientific achievements were put on display, to be readily compared and held against one another. Standardization also meant imposing one nation's measuring system or units as the newly binding standard for all.

Internationalism often resulted in the founding of international bureaus that were home to international organizations but situated in one country alone. Among the earliest such organizations was the International Telegraph Union with a seat in Berne (est. 1865). Other prominent examples were the Universal Postal Union (Berne, 1874), and the International Statistical Institute (The Hague, 1885). In the heated climate of the 1910s, Franco-German time politics eventually clashed over the establishment of such an international bureau. After "losing" the prime meridian battle to Britain, French hopes rested on gaining a lead in international radiotelegraphy, a technology that was being developed and tested in the closing decade of the nineteenth century and the early years of the twentieth. Radiotelegraphic experiments had been conducted at the Eiffel Tower under the aegis of the Bureau of Longitudes since roughly 1900. By the end of the decade, the Bureau expressed the goal to establish a wireless time signal service with the tower functioning as its main antenna. The service began to operate in May 1910 and henceforth emitted hourly time signals, at this point mainly to ships and navigators. Encouraged by this success, the Bureau soon aspired to undertake more.[71]

It was the Bureau's shrewd and experienced head, Guillaume Bigourdan, who took the lead. In the past, it had proved difficult to achieve cooperation in timekeeping between national observatories. In 1912, Bigourdan drafted a note to the Ministry of Public Instruction, informing it about the current state of time signal emission in Europe. Besides providing an inventory of European time signal emissions, the message Bigourdan really conveyed was not difficult to parse—namely, that France in general and Paris, with the Eiffel Tower, boasted superior suitability for becoming the center of European radiotelegraphic timekeeping, especially when compared with German facilities. Most important, Bigourdan urged that "it is in France's greatest interest to seize the opportunity as early as possible and propose the gathering of an international conference in Paris to study the unification of time signals." Any further time wasted meant offering other nations a chance to launch their own plans for a timekeeping center.[72]

Rumor had it that Germany was on the verge of proposing its own emission station and that especially the Geodetic Institute in Potsdam near Berlin was planning to hijack global time signal services at the upcoming telegraphy conference in London in the summer of 1912. German scientists had been conducting their own successful experiments with time signal distribution based off the northern German telegraphy station at Norddeich. Following the French initiative, two international "time conferences" convened in Paris in October 1912 and 1913. In what had to be considered a deliberate snub, British astronomers at the Royal Observatory and officials with the government could not have cared less about the French attempt to regain stature in international timekeeping. Britain considered the meeting to be of such negligible importance that it initially refrained from dispatching a delegate to attend. It was only following hasty French attempts at convincing British counterparts of the meeting's centrality that British delegates eventually arrived on scene a few days into the first conference. The meeting in October 1912, attended primarily by nondiplomatic personnel, passed a resolution for the establishment of a timekeeping bureau located in Paris.[73]

At the two time conferences held in Paris in 1912 and 1913, Franco-German rivalries came to the fore. While contrary to French anxieties Germany did not launch a plan for a German bureau, German officials took issue with several of the stipulations the first conference had adopted. Representatives of the German ministries involved in timekeeping attempted to limit the time span covered by the convention. As one representative put it, the current superiority of French institutions for time-signaling might not last, and in that case, a relocation of the time bureau to an institution better equipped for such services would be desirable. That such an institution would likely be the International Geodetic Institute near Berlin was a thinly veiled conviction among German bureaucrats.[74] The German representative at the conference labored doggedly to keep the convention text from naming Paris as the seat of the international bureau, even though this was where the new institution would open shop according to previous deliberations. France plainly rejected these German demands for amending the convention text and punted on the question of location.[75]

The next target of Franco-German animosities was the Eiffel Tower. Since the justification of choosing Paris as the site of the Bureau had been the availability of an exceptionally well-suited emission station, the French government would have to guarantee that it would always have access to the tower for the purpose of signal emissions. Since the tower was privately owned, such a guarantee could not be granted. The German representation at the conference demanded of the French government a guarantee, inserted into the convention text, that would ensure the availability of a extraordinary emission station at all times, even if the Eiffel Tower service for any reason would have to be discontinued.[76] Throughout these discussions, internationalism appeared more charged with nationalist sentiments than ever. It fit the picture, then, that the United States only agreed to sign the convention under the condition that it would continue to emit its own time signals anyway. The US Navy had begun sending time signals in 1905 and was preparing to set up new stations in Panama, Hawaii, and Samoa. National and imperial concerns placed international cooperation firmly in the back seat.[77]

Together, the conferences of 1912 and 1913 drafted a convention to set up the International Association for Time. Among the organs that made up the association was the International Bureau for Time, which, Franco-German squabbles notwithstanding, was to take its headquarters in Paris. In the words of the French Bureau of Longitudes, the conference outcome was lauded as an "important scientific success for our country." Unwittingly, France was running out of time. Before the project of an international bureau could be completed, Europe was skidding into a conflict that upended the world as contemporaries knew it. In 1919, the International Timekeeping Bureau, still at the Paris Observatory, was realized in a heavily downsized version and became part of the International Astronomic Union, founded in Brussels in 1919.[78]

After the Paris time conference came to a close in the fall of 1913, it was in the context of war and following occupation that German and French national times once again became the subject of disputes. When French troops held and occupied the Rhineland in the west of Germany after four years of bloodshed had come to an end in November 1918, French time arrived in tow. French authorities replaced Germany's Central European Time in the occupied territories with France's own mean time one hour slow. According to the German side, the right to set the time had fallen back to Germany in the occupied western territories with the conclusion of the Peace Treaty at Versailles in June 1919. Once the treaty entered into force, German time should be reinstated, they demanded.[79] The Interallied High Commission for the Rhineland, which took control of the Rhineland once the treaty became law in early 1920, instated a mixed solution: French time would be kept on railways and German time would be kept in civilian life. France soon thereafter switched to summer time, advancing clocks one hour. Temporarily, therefore, French time and German time in the occupied western German territories were identical. Chambers of commerce, artisanal chambers, clubs and associations, agricultural workers, and workers in industry, trade, and craft all took to protesting the violation of German temporal sovereignty.[80] Once the end of the summer time period approached, German authorities instructed diplomatic representatives in Paris to pursue the revision of French time "emphatically." Should such efforts prove futile, German time would be nonetheless kept in civil life as an act of protest and assertion, despite the difficulties arising for railway timetables and traffic from maintaining two times in the occupied zone. "Political reasons" trumped logistics in this case, and transportation had to "step back behind the national point of view," as the Foreign Office in Berlin phrased it.[81]

French and German time politics demonstrated how the adoption of nationwide mean times was conceived in regional and national terms, to eventually become the site of nationalist contestation. But despite such ideological struggles over mean times, the politics of nationalizing time were rarely straightforward. When legislators and administrators discussed the abolishment of a multitude of local times in favor of one national mean time, their thoughts evolved around questions of regional railways and the facilitation of nationwide communication and bureaucracy. Often, as seen in both France and Germany, local authorities such as municipalities unilaterally pressed ahead with time changes without awaiting a decision at the center. They thus forced national authorities to act by sanctioning individual initiatives already in place. Building nationally uniform times therefore appears as a seesawing movement of measures taken by local, regional, and national authorities often in close temporal sequence or simultaneously. Cumulatively, such steps ended with the establishment of uniform times. The implementation of nationwide mean times thus contributed to and was part of state-building. The exchange of ideas and knowledge about time and timekeeping, propelled by a globalizing world and mediated by reformers and countless cheapened pamphlets and brochures, resulted in the fortification of states and the delineation of regions, not least because much of the knowledge that circulated in the nineteenth-century world was or was construed to be *about* the state and the nation. The implementation of mean times in Germany and France also suggests that time and time zones were more than yet another internationalist movement that went "global" in the second half of the nineteenth century, such as the women's movement, the metric system, or international law. Germany's Central European Time and France's resistance against British time indicate that time served to ground national times and territories in global time and space.

CHAPTER TWO

# Saving Social Time

SOME TWENTY YEARS after a number of European states had discussed and eventually attempted to apply nationwide mean times, time in another, less legal, more societal role began to worry self-declared experts, reformers, and government officials. Clock time now took on social rather than bureaucratic, rationalizing meaning. The rational time of ordered national and regional space, of legal and train time, gave way to questions about the useful deployment and division of society's times—work time, leisure, and recreation. Every day during the summer months, reformers now deplored, precious daytime was lost when vast parts of society lay sound asleep while the sun was already up. Time that could and ought to be spent on useful activities thus went to waste. Inadvertently, these wide-ranging debates about social time turned into a display of time consciousness and interpretations about the nature and meaning of time. In the process, discussions about what would become so-called summer or daylight saving time revealed how thinly spread the actual public display and distribution of accurate mechanical clock time had remained after national governments had adopted mean times.

Historians have long assumed a connection between the proliferation of mechanical clocks, the understanding of abstract time stripped of natural qualities, the rigor and routines of workshop and factory labor, and the emergence of a time consciousness and internalized time discipline. They are divided over what came first—inventiveness and an interest in time that gave birth first to the development of mechanical clocks, then to time orientation and discipline, or the demands of regular working hours that generated a demand for clocks and, especially, improved technologies only in the second place.[1] The most influential thesis associated with the question of material clock time, time consciousness, and capitalism is still that advanced in E. P. Thompson's article "Time, Work Discipline, and Industrial Capitalism." Thompson argued that some time in the late eighteenth and early nineteenth century, the increased regularity of factory work led people to abandon what he referred to as "task orientation," the "organization of time according to the necessity of performing particular tasks," mostly in disregard of the time spent in carrying out these tasks. Under task time, work was irregular; the performance of certain labors such as cutting the corn during harvest or milking cows was followed by periods of less intense work or even idleness until another bout of activity was dictated by the rhythms of nature.[2]

As Thompson argued, when more and more people left agricultural work behind and toiled in workshops and factories under overseers equipped with clocks and watches, task time gave way to an internalized time orientation, a consciousness that understood time to be the monetized time of uniform working hours. Puritan ethics also did their part in preaching time thrift. Time no longer simply passed but was now a currency that could be spent. In such an interpretation, there was an implied relationship between the external time of proliferating mechanical timekeepers, the synchronized work rhythms of industrial capitalism, and the internal, intellectual, and mental capacity of ordinary men to lead a "time-oriented" life, fully conscious of the value of time. Thompson does not explicitly mention "abstract" time, and as Moishe Postone explains, the juxtaposition of "task time" with "time-sense" in Thompson is mostly but not entirely congruent with Postone's own "concrete" and "abstract" time. But Postone and others who followed in Thompson's footsteps have read the adoption of mean times that replaced the time of the sun in the mid- to late nineteenth century as a further extension and refinement of this move away from task time to what they now termed abstract time.[3] If Thompson and those elaborating on his arguments were right, by the beginning of the twentieth century the rhythms of daylight and darkness, the sun and the seasons, the dews of the vanishing summer, would have long since ceased to bear meaning in people's "time-sense."[4] But as the history of daylight saving suggests, the connection between work rhythms, time consciousness or the ability to imagine abstract time, and the external mechanical time of clocks was more nuanced than a host of older works suggest and developed along an entirely different chronology.

Rather, around the globe, a multitude of articulations of time coexisted at the turn of the twentieth century, most of which moved along informal, often lateral trajectories instead of trickling down from global to local contexts. Even within single countries, the spread of abstract mean times remained diffuse, and top-down implementations of time laws proved difficult to execute and slow to affect social behavior and mentalities alike. Instead, here too, the simultaneous presence of multiple time regimes dominated. Whether in Germany, France, the United States, or Britain, solar time and the time of other natural rhythms still shaped the social imagination of time, even in an age of cheapened pocket watches, proliferating public clocks, and clearly delineated (and contested) working hours. A swelling working class and multiplying white-collar work force, imbued with the temporal discipline required by factories and shop hours, did not bring with it a wholesale disappearance of natural rhythms either. Industrial capitalism appears to have benefited from notions of time thrift and time discipline. It did not, however, require abstract time, and not even widely available accurate time, to conquer the globe and reap unprecedented profits.

• • • •

IN 1907, the British builder and inventor William Willett published a pamphlet titled *The Waste of Daylight*. Born in 1856, Willett had joined his father's construction business earlier but around 1900 began to focus his energy solely on promoting the cause of daylight saving. In his essay, Willett raised several issues that would continue to dominate the discussion for the decades to come. "Everyone appreciates the long light evenings. Everyone laments their shrinkage as Autumn approaches," Willett surmised. Nevertheless, he continued, "standard time remains so fixed, that for nearly half the year the sun shines upon the land for several hours each day, while we are asleep, and is rapidly nearing the horizon when we reach home after the work of the day is over. Under the most favorable circumstances, there then remains only a brief spell of declining daylight in which to spend the short period of leisure at our disposal." Willett desired to resolve this conundrum with a simple trick: "Now, if some of the hours of wasted sunlight, could be withdrawn from the beginning, and added to the end of the day, how many advantages would be gained by all." To Willett, the primary objective of saving daylight was motivated by public health concerns: his goal was to increase "existing opportunities for exercise and recreation." Precious daytime would thus no longer be wasted sleeping but utilized for useful activities.[5]

Willett had studied time and daylight saving time for several years when he authored the little booklet that would prove so popular far beyond Britain. He understood that uniform mean time was an abstraction of sorts, and that summer time was simply an even more abstract move designed to outsmart nature's distribution of daylight over the course of twenty-four hours. Under summer time, 7:00 a.m. would become 8:00 a.m., forcing people to change their rising time in relation to sunrise time. Work would finish accordingly earlier, leaving more time between the end of the workday and the fall of darkness. When Willett stated that uniform mean time remained "so fixed," he imagined a society in which the availability of stable public time was ample, in which his contemporaries had internalized and comprehended the abstract mechanical time of public and personal timekeepers and the regiment of daily work schedules.

At the beginning of the twentieth century, Willett and an army of social and public hygienists, medical experts, sociologists, and an increasingly time-intrigued public discussed intensively the social use and meanings of time in conjunction with other social issues. Concerns about human fatigue and the waste of energy gave rise to Taylorism and its attempts to improve the efficiency of factory work. Questions of work and leisure, of achieving a healthy balance between the two, had become major preoccupations in mature capitalist societies. Since the onset of the Industrial Revolution in Britain, the percentage of the population that was working in agriculture declined dramatically, by one estimate from roughly 35 percent in 1801 to only 8 percent in 1901.[6] At the same time, manufacturing, and the service industries and public sector, occupied a rising share of the economy. In Britain, this shift of output away from agriculture to manufacturing, industry, and services was accompanied by the growth and diversification of the working- and middle-class professions, and the emergence of a white-collar work force from the 1880s.[7] These transformations left deep traces on people's understanding of time. Since British industrialization had been firmly under way for more than a century, debates about daylight saving and the social meaning of time were particularly pronounced in Britain. The struggle over summer time thus reveals much more about European imaginations and conceptions of sociotemporal rhythms than is commonly acknowledged.

The use of daylight had been broached when time zones were first introduced. In Germany and elsewhere, a difference of more than fifteen minutes between solar time and mean time was taken to be considerable. Differences of up to thirty minutes, a common result of the adoption of mean times in the eastern- and westernmost parts of many countries, were regarded as almost too much to handle. In order to hew as closely as possible to the rhythms of light and darkness, many nineteenth-century institutions informally followed a practice that mirrored the idea behind daylight saving. During the summer months, school and work started earlier than in the winter. In the first decade of the twentieth century, such practices suddenly found a growing group of followers as daylight saving associations and clubs sprang up in North America and beyond. In the United States, the

National Daylight Association scored a victory in 1910 when Cincinnati was among the first American cities to advance the clock by one hour during the summer months. In the Australian state of Victoria, a summer time bill was discussed in parliament in 1909; a similar conversation took place in New Zealand. Even prior to that time, in parts of Australia and in the British Cape Colony informal summer time arrangements had been put in place in the 1890s, advancing time some twenty (Australia) or sixteen (Cape Colony) minutes in the summer to arrive at a better distribution of daylight throughout the day.[8]

It was William Willett's pamphlet that got debates truly under way. One year after its publication, in 1908, Willett's plan for saving daylight was incorporated into a bill that came before the British House of Commons at the initiative of Member of Parliament (MP) Robert Pearce. After much discussion, summer time was for now stayed without further steps taken.[9] In 1909, another MP presented a new bill, which ended up before a study committee. After further investigation, the committee found that daylight saving plans should be abandoned due to the serious objections and contradictory evidence the committee had unearthed. With opinions divided, daylight saving bills were introduced in 1911, 1912–1913, and 1914 but in no case advanced beyond initial readings in parliament. Members of the British government such as the Home Secretary feared insufficient support for a summer time bill to make it through parliament. "So long as the Bill is not generally understood, [and] its real benefits are not appreciated by the large masses of the public," the government's hands were tied.[10]

It required the exceptional circumstances of World War I for daylight saving to become reality. In the spring of 1916, MP Henry Norman moved in the British parliament to reconsider daylight saving as a means to shorten hours of artificial lighting and to economize coal, oil, and other fuels in times of war. Only one month earlier, Germany had been the first country to adopt daylight saving as a wartime measure to conserve energy. The fear of granting the German opponent an advantage met with the general perception that "wartime" called for drastic measures, and eventually opinions among British government officials tipped in favor of daylight saving. The summer time bill was passed in May 1916.[11] To prepare for a possible repeat of daylight saving in the coming years, a study committee was charged with exploring the effects of daylight saving after the first year's experience. The committee canvassed a broad spectrum of the population and mailed questionnaires to authorities, interest groups and professional organizations, and individuals, receiving roughly 1,300 replies. A reintroduction of summer time was recommended for 1917 and subsequent years.

The 1916 act was extended several times up to 1922 under a policy that allowed the continuation of war emergency laws. But in 1922 those special measures had expired, and for the first time since 1914, a parliamentary debate addressed daylight saving when a bill proposing the extension of the measure was brought before the House of Commons. The bill would have given summer time a permanent statutory existence instead of debating it anew every year and fixed the time period during which daylight saving would be in force. Now that the war was over, however, summer time no longer had the support it enjoyed during the war years. The new bill that eventually became law did not make summer time permanent, and that meant new discussions every year. From 1922, each spring, the same arguments for and against summer time were aired, the period of application slightly amended until in 1925, daylight saving in Britain was fixed for some time. These repeated contestations testified to the deeply divided opinions on summer time among the public and government officials alike.[12]

In Ireland, summer time became a welcome opportunity for British authorities to not only put clocks one hour forward and back again but to introduce Greenwich time. The desire to bring the unruly island into the British fold had been first expressed in 1898 in a memorandum composed by the Royal Scottish Geographical Society that was sent to various government agencies. Back then, administrators at Britain's General Post Office declared the matter outside its purview, deferring to other agencies to broach a potentially divisive issue. The Post Office would be "overwhelmed with communications" it was unable to cope with.[13] The question of Irish time does not appear to have been discussed much further until in 1911, when France finally adopted the time of the Greenwich meridian. Suddenly it was surmised that the other unruly opponent of "British" universal time, Ireland, might be convinced to follow in its steps. Some Irish commentators immediately condemned Greenwich time as an "instance of Saxon tyranny."[14] As several decades earlier, when France rejected "British" time, the political qualities of mean times as indicators of geopolitical affiliation could hardly be concealed.

Later in the year it became known that Portugal would soon observe Greenwich time as well, once more prompting British excitement about the impending extension of "British" time to obstructive Ireland. Now that "almost every country in Western Europe" had adopted Greenwich time, the use of Dublin time throughout Ireland was an anomaly. "Uniformity is one of the considerations of the moment," a member of the House of Lords found.[15] In the summer of 1912, the so-called Uniform Time Bill proposed to extend the application of Greenwich time to Ireland.[16]

There was, as the sponsor of the bill acknowledged, a certain amount of utopianism in a proposal to make Ireland follow British time at a moment when granting Ireland a parliament of its own was a seriously discussed option and when Irish calls for more autonomy were growing louder.[17]

The "Uniform Time Bill" for Ireland passed the House of Lords where it had been introduced but was never considered by the House of Commons. With the outbreak of the war approaching, plans to introduce Greenwich time in Ireland only resurfaced in 1916 in connection with the passing of summer time laws. Simply introducing summer time and Greenwich time at the stroke of one law would reduce inconveniences and finally effect the "unification of British and Irish time."[18] In the fall of 1916, Irish railway stations, post offices, and police barracks were covered in thousands of posters that had been printed to announce Ireland's move to standard time. In bold capital letters, these announcements read "TIME (Ireland) ACT, 1916. On and after SUNDAY, the 1st October, 1916, Western European Time will be observed throughout Ireland. All clocks and watches should be put back 35 minutes during the night 30th September–1st October." This was a generous reading characteristic of an age in which seconds still did not necessarily carry much practical relevance, as Dublin time was actually thirty-four minutes and thirty-nine seconds behind. As in France, the "British" provenience of the new time was slyly obscured behind the seemingly innocuous title of "Western European Time." If the "Central European" zone had expressed German and, to a lesser degree, Austro-Hungarian geopolitical realities and aspirations on the continent, the "Western European" zone was to detract from the nationalism inherent in time politics. To little avail. A year later in 1917, several Irish vicars were said to have refused advancing the village clock one hour. National identities followed universal time wherever it went.[19]

German experiences with summer time during and after World War I followed a similar path as those of Britain. The country had been the first in Europe to adopt daylight saving in 1916, but a prolonged volatile political situation precluded a statutory solution after the war. During the war, the government could decree summer time, but with the expiration of such extraordinary wartime powers, the struggle for public opinion was fully ignited. In 1919, the German parliament took to discussing daylight saving for the first time. During these debates, several opponents of summer time cited the explosive political situation in the country as driving their opposition. Germany had just the month before in the fall of 1918 witnessed a revolution, and in several regions of the new Weimar Republic, the government continued to be under attack from both right- and left-wing forces. If, as evidenced, miners in all parts of Germany strongly opposed summer time, and if railway workers even threatened to go on strike, daylight saving constituted a dangerous addition to an already combustive political blend.[20] The summer time draft bill of 1919 was rejected in parliament. In 1924, renewed scrutiny of the summer time question by the ministry of the interior was silenced with the remark that agitation was rising in preparation for the upcoming national election campaign of the same year, and that despised topics like summer time should not be thrown into the restive mix of electoral politics.[21]

France was no different from Britain and Germany. Here, the summer time bill was debated in the Chamber of Deputies in April 1916 and subsequently passed on to a commission.[22] In the end, it was the prospect of fighting a war alongside allies who had adopted summer time that convinced the commission. National defense temporarily trumped all other concerns.[23] After the war, the French Chamber of Deputies voted in favor of permanent summer time by a narrow margin only to abrogate the law in 1922 after ongoing protests and criticism. The French law even went as far as to allow the prefects of rural districts to set opening and closing hours of schools, fairs, and markets as well as railway timetables in such a way that ensured conformity with the necessities of rural life. Certain localities could thus opt out of summer time, a loophole that effectively guaranteed an uneven and incomplete application of daylight saving throughout France.[24] As in Britain and Germany, debating daylight saving time was now an annual ritual.

As with standard time, summer time prompted efforts at international coordination. Such cooperation was driven simultaneously by western European nation-states and their interests and international organizations. As a report by the British Home Office complained, the "absence of uniformity" in the duration of summer time was a source of inconvenience especially for the international transportation of goods and passengers. As early as 1921 and 1922, Belgium, Britain, and France strove to reach an agreement over a uniform period of summer time, but different interests in the participating countries did not see eye to eye. British and especially Scottish agricultural associations opposed any significant extension of summer time into October, whereas in France, legislators, the tourism industry, and transportation officials opposed an end to summer time during the holiday month of September.[25] On the international level, plans to coordinate summer time emanated from the League of Nations. In 1922, the secretariat of the League wrote to European governments to encourage the international standardization of summer time start and end dates. Around the same time, the International Union of Railways passed a resolution calling for governments to reach an agreement on the beginning and end of annual daylight saving. The preparation of timetables especially for continent-wide traffic happened with significant advance planning, whereas individual governments announced their annual summer time policies on rather short notice. Europe-wide uniformity of daylight saving was not realized until in 1998, when the European Union finalized a number of directives that fixed start and end dates permanently for member states.[26]

The arduous process of negotiations in Britain, Germany, and France, the repeated criticism and following reversals in each of these countries, and the inability to agree on a common duration for daylight saving indicate how contested summer time was on the political level alone. Once officials realized summer time's social potential, antagonism only grew. Opposition to summer time was stronger than it had been to standardized mean times. In part, this was because a general interest in time had increased over the course of a few decades. The general public was now fascinated with the supply of accurate, uniform time. Discussions about daylight saving therefore attracted more attention and, accordingly, expressions of discontent. Other reasons had more to do with the specific impact of daylight saving on people's social lives and their understandings of the essence of time. From the first decade of the twentieth century through the interwar period, interpretations of the nature and social function of time were still in flux.

• • • •

DAYLIGHT SAVING was decades in the making. Attempts at passing legislation were interspersed with tense debates. A permanent regulation remained out of reach, as daylight saving was called into question time and again. No consent was forthcoming even on the apparently procedural question of whether legislation was indeed the appropriate step to take in regulating people's habits of time use. Such doubts are a forceful reminder that time was not like any other matter to become the subject of a law. The symbolic and highly personal and individual qualities of experiencing time rendered it different from other areas touched by legislation or internationalized and standardized as part of nineteenth-century globalization. Daylight saving bills were occasionally criticized as an instance of undue government intervention into the private, social lives of individuals. Rather than regulating human behavior by law, such critiques argued, it should be left to private businesses to arrange for an earlier beginning of the workday in order to achieve the same goal. A reader writing to the British newspaper *The Spectator* complained that William Willett's daylight saving plan "proposes to put us to bed and get us up by Act of Parliament. Personally, I like to choose my own time for these operations." Voices that opposed an act of legislation as a nudge to encourage early rising moreover accused those in need of a law of being "a few bloated millionaires who cannot be got out of bed," or "a rich and influential faction with a hobby." Hardworking ordinary people, it was implied, would need no such coaxing.[27]

Those who did advocate for summer time couched their arguments in the language of various strands of social reform. From this perspective, laws about time were yet another tool for "correcting" human behavior and for steering and engineering society into the desired direction and shape. As one such example, daylight saving was proclaimed to aid the cause of temperance. Pubs were frequented primarily after dark, and extending daylight would therefore leave the lower classes with less time to indulge in their habitual heavy drinking after work. Another favorite topic among social hygienists and public health advocates was neuralgic health and nervous diseases. Proponents of daylight saving argued that "the nervous strain in the young" was increasing and that the "only real antidote" was fresh air, to be enjoyed in greater measures after school under a summer time regime.[28]

Beginning in the second half of the nineteenth century, a combination of trade-union activism, half-day holiday movements, and government legislation gradually reduced working hours and prescribed breaks from work during the week. The six-day workweek and even Saturday half-holidays became more common in many industries. Consumption and leisure were moreover enabled by rising real wages, providing more sections of the working and middle classes with spare income to spend on entertainment.[29] Since the late nineteenth century, mass commercialized leisure became a reality even for workers through institutions such as clubs and associations, music halls, and modest railway excursions. Middle-class Britons, meanwhile, spent their free time in seaside resorts, took a walk in parks or public gardens, shopped at department stores, and participated in a growing number of athletic activities such as golf, tennis, and cricket.[30] The clear separation of work from leisure fostered attempts among social reformers of a different shade to goad workers into indulging in "rational recreation" activities and to guide them away from public houses and other sites of numbing idleness and temptation. It was not by coincidence that the leisure habits of working-class males in particular became a target of reformism in an age when two laws in 1867 and 1884 extended the franchise to include a growing number of working-class men.[31]

The reasons behind encouraging workers' recreation were manifold and ranged from conservative social-reformist ideas about improvement, temperance, and family values to political goals of feeding working-class voters useful knowledge, to philanthropy and concerns about health and productivity at the workplace. Such variegated motives notwithstanding, as ideas about workers' welfare and recreation gripped the imagination of politicians and analysts of social life, those claiming to speak on behalf of workers' interests were adamant about the recreational and medical benefits of daylight saving for members of the working classes: "We know the toll that is taken of our health in the towns through the absence of fresh air and sunlight," one member of the British parliament warned. "Therefore, the voice of the teeming millions who live in the towns, the horny-handed sons of toil who work in the shipyards, blast furnaces and iron works is practically unanimous in favor" of summer time, he posited.[32] Yet leisure and consumption were even more obviously privileges of the middle classes. The most frequently referenced activities to be enjoyed during the newly gained free time were cricket, tennis, and golf, middle-class pastimes par excellence. Critics therefore often cautioned that the true beneficiaries of summer time were the urban middle classes rather than workers.[33]

Ideas about public health and social hygiene and the social times attached to them were infused with notions of race. A representative of the Women's Amateur Athletic Association underscored that summer time would increase opportunities for physical recreation, and "the need of our Age is to provide healthy open air exercise for the young women who will be the mothers of the next generation," one Ms. Cambridge held forth in front of a committee that was studying daylight saving. "Summer Time gives these girls, on whom we depend for a strong healthy future race, an extra hour during which to get away from the factory."[34] A retired professor speaking out in favor of daylight saving similarly declared the health benefits of playing open-air games to be crucial to the "future of the race." William Willett himself had gone so far as mentioning the health benefits of summer time with regard to a possible war when he wrote, "Against our ever-besieging enemy, disease, light and fresh air act as guards in our defense, and when the conflict is close, supply us with most effective weapons with which to overcome the invader."[35]

Opponents of summer time, too, looked to public health and social hygiene when making their case. The committee that had been appointed to study the effects of summer time in Britain in 1916 worked hard to address the serious accusation that summer time would be detrimental to the health of children. As critics held, summer time stood in danger of curtailing sleeping hours for those most in need of a sound night's sleep, for neglectful parents, to be found especially among the working classes, would surely allow children to stay out in the streets beyond their proper bedtime.

According to the study committee's inquiries conducted after 1916, a number of districts indeed reported evidence of tiredness and lack of attention among schoolchildren after the first year of British daylight saving. The committee hastened to explain that negative effects were few, and that in those cases where fatigue had been observed, sleepiness among children was mostly attributable to a lack of discipline in poor households where conditions were "lax" when it came to child-appropriate bedtimes. In a similar vein, many authorities commented critically on the "carelessness" of parents with regard to bedtime hours. In an effort to tighten the surveillance of such neglectful behavior and to monitor the social times of families, teachers had been instructed to report on a lack of attention and unusual sleepiness so that the matter could be taken up at parent-teacher conferences.

In an age obsessed with demographic scenarios of population decline and the need to maintain steady numbers of births and deaths and a healthy race, children, the future of the nation and the race, were considered too precious to be left to the lenience of lower-class families. This interest in health and race notwithstanding, women and "national motherhood" were absent from the controversies surrounding daylight saving and time reform. Notions of gendered private and domestic time had already been transformed as more women entered the workforce in the second half of the nineteenth century. During the war years gender roles underwent even more marked transformations when women replaced men who had been drafted at industrial workplaces central to the war effort, eventually affording women the right to vote in many countries in the years thereafter. Yet female time never entered discussions about time reform, not least because the self-styled reformers who argued for "rational" and accurate time were all male and preached values widely viewed as a male prerogative.[36]

• • • •

THOSE OPPOSED to summer time clad their critiques in a variety of arguments. A citizen writing to the Home Office in 1916 complained about the failure to consult people's opinions before adopting such far-reaching policies. Changing "true local time" was "socially objectionable as it is of the nature of servile government," he wrote. "We have had no chance to object to it by one vote," and "I object to pushing a measure like this thro' Parliament at short notice, surely the people should have something to say about compulsory early rising!!!!!"[37] Others worried less about the democratic nature of the decision to observe summer time and instead concentrated their criticism on the workings of international finance capital. Prior to summer time, the last hour during which the London Stock Exchange was open (3:00–4:00 p.m.) was the first hour of business at the New York Stock Exchange. Prices from New York were telegraphed to London at around 3:00 p.m. Daylight saving would cut off the period of overlap during which a large amount of business was conducted. Often, business continued in the streets for a while after the London Stock Exchange had officially closed its gates.[38]

Still, the most vociferous and lasting resistance against summer time did not emanate from worried nutritional experts, public hygienists, or stockbrokers, but rather from agricultural workers and organized agrarian interests. The debates about summer time were implicitly debates about what future there was for agriculture in a country that was rapidly industrializing. While in other Western European countries agriculture continued to employ between 30 percent and 45 percent of workers, in Britain these numbers had dropped to the single digits.[39] It was claimed that industrial workers like the Scottish miners would benefit from summer time since it stood to lengthen their day beyond work underground. But agricultural workers continued to toil in sync with the rhythms of nature, and nature would be at odds with an advanced clock. Organized agrarian interests like the Scottish Farmers Union and the National Farmers Union of England and Wales were at the forefront of protesting summer time from its beginnings around 1907 all the way until 1925, when summer time became permanent. The government and the Home Secretary in particular were well aware of the mood among farm workers and on several occasions refused to go forward with a bill or merely declaring their support for a bill, citing a lack of support and "strong opposition" in agricultural districts.[40]

Summer time forced those who had to rise early to rise in the dark and to abandon time otherwise spent over breakfast with the family before making the commute to work. Certain unalterable rhythms were understood to govern the rising times of agricultural workers. On the fringes of summer time in April and September, farmers rarely began work before 10:00 a.m. as heavy dew took its time to dry off. To make up for the late start, during the hay and grain harvest, farm workers normally worked later in the evening. Under daylight saving, these workers would be precluded from enjoying any recreational time after work.[41] In one parliamentary debate on summer time, Sir Henry Cautley of East Grinstead rose to take the side of those disadvantaged by summer time. The early risers supplied the "wants of the townsman," he charged. Among them were domestic servants who had to catch early trains; wives preparing breakfast for their husbands; railway men, steel smelters, tinplate workers, and men in large factories of the north.[42] "Cowmen and horsemen must be in the stalls by 5:00 to 5:30 to get milking done and drive three to four miles to catch the first train to send milk all over the country. Farmers and their men work three hours and a half before shopkeepers think of opening shop," Cautley warned.[43] For all these people, summer time made it necessary to rise in the dark for at least one additional month every year. Farmers only began to accept the idea of daylight saving in the mid-1920s when, "in face of the overwhelming opinion of the urban community," it became clear that it was "useless to object to summer time as an institution." From now on, farmers focused their energy on amending the time span of its application.[44]

In addition to the rhythms of farm workers, another element of rural life was thought to stand in the way of summer time: the presumed patterns of bovine behavior. The cow and her proclivity to give milk at a certain hour of the morning became one of the most discussed aspects of the politics of daylight saving time. It was not uncommon for parliamentary and other discussions to converse about the cow as the "chief opponent" of summer time, about "her habits" standing in the way of a time change, and about how summer time would "disturb the habits of the cow."[45] Others refused to believe that the cow would change her habits. "Cows are creatures of more regular habits than possibly the men who attend upon them," the 1909 report on summer time ventured. "Cows know to an hour when milking time is come and when it has not," hence cows would not be ready to give their milk at the new time. Focusing on the cow, together with the perceived backwardness of rural society, was often an exercise in mockery. Cows should not seriously "set the limits of summer time," one newspaper reader found.[46]

Another frequently voiced unease with summer time stemmed from its potential to disrupt established protections against extended working hours. By the turn of the twentieth century, in countries such as Britain, France, and Germany the length of the workday had become the subject of at times fierce struggles between organized labor and employers, between government regulation and inspection. A number of acts curtailed the working hours of women and children. Early-shop-closing movements fought for the reduction of opening hours as a means to reduce the working hours of sales clerks and employees in the retail and leisure trades; at twelve hours, their work time often even exceeded the hours of production workers.[47] One popular instrument to safeguard against long working hours among this section of the working population was closing orders, requiring shops to close normally no later than 7:00 or 8:00 p.m. Those harboring doubts about the benefits of daylight saving regularly mentioned the possibility of extended opening hours resulting from changing the clock and gaining daylight. In 1916, the National Amalgamated Union of Shop Assistants, Warehousemen, and Clerks wrote to the British Home Office with a request. Was it possible, the union asked, to promote the adoption of a "Closing Order" as it was common elsewhere, in all those towns and districts where such an order was currently not in force? It was to be expected, the union explained, that shopkeepers would be tempted to keep shops open longer as a corollary of summer time. Shop assistants would therefore be cut off from enjoying recreation and fresh air after work.[48]

Such worries may not have been entirely unfounded. In the same year of 1916, a branch of the National Association of Goldsmiths in Cheltenham addressed the Home Office with a request to be allowed to use daylight saving time to "evade" the Shops Hours Act of 1904. That act had set the closing hour for 8:00 p.m. Now that daylight saving had been adopted, the jewelers asked, would it not be possible to keep shops open until nine, since that was in reality only eight o'clock? The Cheltenham jewelers were politely informed that the existing closing act would have to be read according to summer time, and that nine o'clock was therefore nine o'clock. Soon after the initial adoption of summer time in Britain in 1916, reports about summer time "abuses" surfaced. The "better class drapery firms" in London's West End were known for keeping open thirty minutes to one hour after official closing time. A small draper's shop in North London had announced to employees that shop hours would be extended to nine o'clock now that it was light at that hour, thus "marching out of time," as one paper put it. The Home Office decided to instruct its shop inspectors to keep an eye on the situation.[49] Much later in the final decades of the twentieth century, the decline of manufacturing and the rise of the so-called service industry again transformed the meaning of work time and leisure. In some ways, time has been unstandardized since the 1970s. The permanency of electronic communications and financial transactions has rendered the concept of time zones obsolete to some industries, expressed in the notion 24/7. Working hours have moreover become "flexible," including work from home, and thus all but blurring the lines between activity and rest, work and recreation. In the face of these developments some have questioned whether overall, the length of the workday has truly decreased since the nineteenth century.[50]

• • • •

ARGUMENTS ABOUT work times, industrial work, farm labor, and leisure reflected the social attitudes toward time in a society that by the end of the nineteenth century held fairly evolved and popularized views on work, entertainment, and public hygiene. Besides the collective times of society, many commenters spoke to the ways in which individual behavior would respond to time changes. What is extraordinary about these remarks is the skepticism and conservatism with which most people viewed time changes even as late as the 1910s and 1920s. Today most people living in a typical one-hour-wide time zone have probably lost any apprehension of how much standard time diverges from sun time. Differences of up to half an hour at the easternmost or westernmost extension of a time zone are common in many European countries and the United States, but few people know and even fewer care. In the late nineteenth and early twentieth century, great uncertainty prevailed as to whether the general public would in fact follow any legal time change that stood to widen the cleavage between mean time and "true time." Rural life and work in particular were understood to simply ignore the change in official clock time. Skeptics held that "whatever might be done with the clocks, agricultural and pastoral operations would still be regulated by the sun," and "any radical change in the English way of living cannot be expected."[51]

George Darwin, son of Charles and a retired professor of astronomy at the University of Cambridge, commented on the proposed time change: "A committee of the House of Commons has now come to the conclusion that it would please every one in the country so much to call 11 o'clock in the summer months by the name of 12 o'clock that the change ought to be adopted." People would be unwilling to adopt the change and continue to use Greenwich time, thus clocks would need two hour-hands, and it was "absurd to suppose that the House of Commons can, by a mere verbal artifice, change the habits of the whole community." Darwin continued, "We shall have to write our times in duplicate like Russian dates in the old and the new style."[52] A concerned member of the public painted another worrisome scenario: "Think of the trouble of altering every clock in the house twice if not oftener each year, and of the results if we forget to do so. How many clerks and shop assistants will turn up late for their work in the morning of the change? How many appointments will fall through owing to forgetfulness of the alteration of the clock? How many trains will be missed that day by those who have neglected to put their watches forward?"[53]

As late as 1916, the symbolic and possibly religious quality and authority that many attributed to solar time was adduced to corroborate such arguments about the impending failure of legally imposed time changes. As a member of parliament recounted, "Many people appear to regard time . . . and our method of reckoning it, as something peculiarly sacred, something irrevocably fixed in the order of things. Based as it is upon the heavenly bodies, it seems to them to possess an almost semi-divine sanction." He went on, "I feel sure there are many worthy people who regard any proposal to tamper, as they say, with the hands of the clock as savoring of irreverence."[54]

Some observers outright demanded that people wearing a watch not set to summer time should be penalized, for otherwise, legal time changes would simply not be obeyed. The 1909 report on daylight saving found that "the question of what amount of pressure would be required to enforce the proposed change does not appear to have received adequate consideration." It surmised, "Nothing would suffice to enforce the change except statutory compulsion under penalties." A reader of the *Times* joked, "One can imagine a future Galileo summoned during the close season for astronomy for willfully exposing a clock showing illegal time."[55] During its first iteration, summer time was in fact ignored in both agricultural and industrial circles, as different European governments soon found out. After the war, French authorities conducted an inquiry into the application and effects of daylight saving and came to the conclusion that workers in a number of industrial sectors and agricultural laborers had been found to force their employers by threat of strike to return to solar time.[56] In the minds of many contemporaries, then, the temporal habits of individuals and society were sticky, and the idea of changing behavior literally overnight and by law was laughable at best and presumptuous at worst.

In tandem with doubts about the feasibility of legally mandated changes of temporal behavior came a germane challenge to imagine time as unchained from light and darkness and physiotemporal rhythms. Most people still strained to think of abstract time. In discussions about the advantages and disadvantages of summer time, many voices continued to picture themselves as living in a world of absolute time that was determined by the course of the sun, which in turn governed the biological rhythms of human and nonhuman life alike. That world and the rhythms of life captive to it were immovable. When mean times were adopted or hours were shifted back and forth under summer time, habits and rhythms remained stable at the same point in absolute time. "It is improbable that persons who dine would alter their dinner hour to conform with Summer Season Time," the British Home Office found in an assessment of 1914.[57]

Sleep and wake cycles were understood to be fixed in time. The scenario painted by summer time cautioners usually saw people and especially families with children rise an hour earlier during daylight saving, but instead of going to bed accordingly earlier, the biological facts of life and, in this case, the rhythms of sleep and nourishment would prevent people from slumbering during an earlier hour in the evening. Instead, dictated by the forces of nature, people would eat and go to bed at the same point in "absolute" time as before. Such inalterable rhythms would disrupt even rail travel. Since the "habits of the traveling public" would remain unchanged, "if the public desired to travel at the same actual time as now, the trains would have to be run at the same actual hours, whatever these hours were called."[58] It would therefore be necessary to change a departure time that had been 7:00 a.m. under mean time to 8:00 a.m., instead of just reading the timetable according to summer time. Heinrich Grone, founder of a commercial college in the northern German city of Hamburg, published a pamphlet advocating summer time in 1916. Speaking about timetables, he noted, "Nothing will have to change in the entire timetable except that all journeys occur one hour earlier. Until a new timetable is published, timetables displayed in train stations will receive in red letters an imprint diagonally printed over the tables: everything one hour earlier."[59]

Some officials were aware of the pitfalls of imagining time and sought to hedge against confusion by anticipating the public's misconception of the shift to summer time. When daylight saving time was first introduced in Britain in 1916, the Home Office took measures to ensure that its instructions were not taken to instate a twofold time regime. On information leaflets distributed to factories it specified that hours normally stipulated as "the commencement and end of the period of employment and meal intervals should *not* [italics in the original] be altered, but they will be read according to the altered time."[60] In Germany, government authorities similarly intervened to prevent confusion over the nature of the time change. Schools in Bavaria and elsewhere had in the absence of summer time in previous years observed informal summer time by starting school one hour earlier in the summer. During the first iteration of summer time in 1916, this practice had been continued even under daylight saving, forcing children to rise in the dark. For the following year, government authorities therefore ordered schools to refrain from applying their own summer time change now that daylight saving was in place.[61]

• • • •

William Willett was right in theory when he wrote of the fixity of Greenwich time as preventing an efficient use of daylight during the summer months. But the frequency with which Europeans stumbled over the real and imaginary hurdles posed by time changes raises doubts as to how stable the unification of time had really been. People wielding less expert knowledge than the self-declared time reformer Willett after all had barely acknowledged the existence of uniform mean time, let alone absorbed its portents for conceptualizing time. In discussions about daylight saving time, there was a remarkable unawareness of previous changes to clock time and, more specifically, the introduction of Greenwich time.

The absence of any active memory or even knowledge of the introduction of Greenwich time is palpable in references to other time changes, measures that served as an illustration of the benefits and disadvantages of adopting a new time. When nineteenth-century British observers compared the switch to summer time to other time changes, they barely ever evidenced the switch to Greenwich time that had occurred in Britain in 1880 and by now should have been firmly established in the daily lives of British society—if Willett was right. It was as if Greenwich time had never been introduced. Instead the public pointed to other cases. One expert whose opinion was recorded in the 1909 committee report stated, "I cannot imagine how they can hold any doubt when they realize that the alteration has been effected without any difficulty in the Cape of Good Hope." Said expert was referring to the introduction of the time two hours ahead of Greenwich in the Cape Colony in 1902. Another witness cited the adoption of standard time in Germany in the early 1890s as an example of an identical transformation that had occurred absent large disturbances.[62] Why was there hardly an awareness of the fact that Britain itself had presumably long since adopted a uniform mean time?

Britain's path toward a countrywide application and display of Greenwich time had begun with a number of infrastructure projects advanced in the first half of the nineteenth century. Railroads forging ahead on a growing number of tracks throughout the country became a familiar sight (and sound) since the 1830s. Electric clocks, soon adorning many train stations, were an innovation of the 1840s. Beginning in the same decade, telegraphy made it easier to send time signals to train stations and post offices along the newly built lines. From the late 1840s, the majority of railway companies were using Greenwich time on their lines. By 1855, in theory, most towns in Britain had succumbed to Greenwich time. At the heart of all timekeeping in Britain was the time determined by the astronomers at the Royal Greenwich Observatory. In the 1850s, then Astronomer Royal George Airy devised a blueprint for disseminating observatory time from Greenwich to other parts of the country. Airy planned for an electronic "master" clock, as it would come to be known, to be installed at Greenwich, fitted in such a way that it would emit electrical impulses in order to drive so-called slave clocks connected to it. Such an impulse could be deployed to ring a bell, drop a time ball, fire a gun, operate a relay, or even set the hands of another clock right. Every hour the clock would send out time signals through telegraph lines running from the observatory to the nearby Lewisham railway station, from where the normal railway telegraph lines would transmit time into London to the central telegraph office. Railways furthermore passed on the signal to all stations where clerks were instructed to regulate platform clocks with the time supplied. In spite of such efforts to distribute the time of Britain's prime observatory widely, local times remained in use on British railways. *Bradshaw's Railway Guide* continued to list the time differences between Greenwich and local times in England even in the 1880s.[63]

Efforts to disseminate time in Britain via networks of distribution multiplied around the same time when governments were discussing the adoption and transmission of similar uniform clock time in Europe and North America. One such system of time dissemination was centered at the London central telegraph office, where an apparatus of almost mythical qualities regulated the onward distribution of time. From the 1850s, the newly opened telegraph office mesmerized the popular imagination of time with its complicated apparatuses and strands of several thousands of wires and powerful batteries. The "labyrinthic abode" of electricity, a room underneath the main floor of the telegraph office, contained a set of special batteries that powered an apparatus called the "Chronopher," often referred to as the "national timekeeper." Each morning when the hands of the oversize electrical clock at the Royal Greenwich Observatory pointed to ten, the Chronopher sent forth a signal, thus connecting stations throughout the country with the Greenwich Observatory at the exact moment of the time signal emission at 10:00 a.m. The Chronopher itself, as well as a handful of other prominent public clocks such as Big Ben's tower clock in London, reported themselves periodically to Greenwich Observatory to have their times checked and corrected. After telegraph companies in Britain were nationalized in 1870, Greenwich time was moreover supplied on an hourly basis to the main provincial post offices across the country.[64]

Since the early days of Greenwich time services, individual or corporate customers such as chronometer makers had the option of renting a private telegraph wire set up to receive the Greenwich signal. When the Post Office took over Britain's telegraph companies it inherited and later expanded existing private contracts, a source of "considerable revenue" derived from these services.[65] The Post Office later had to share the market for private time distribution services with the newly established Standard Time Company (STC). The company received an hourly time signal directly from Greenwich and subsequently passed the time on to subscribers via its own private wires, a service that, albeit offered at a lower rate than the Post Office demanded, was still out of financial reach for the majority of people. Its customers were primarily corporate: the Standard Time Company synchronized clocks for the London Stock Exchange, Lloyds insurance, banks, large commercial houses, and similar businesses.[66] Numerous other companies offered synchronization services, a task that was increasingly sought after by the turn of the nineteenth century. Public electric clocks had become more affordable and widely used, but now they had to be synchronized. Best known in the synchronization business was perhaps the Synchronome Company, which, together with its roughly twenty competitors, sold clock synchronization systems to municipalities, schools, hospitals, and large businesses keen to install the new technology for the surveillance of its workers.[67]

Networks of time and distribution were yet to become stable and wide-ranging. As a consequence, other, less conventional forms of time dissemination persisted. In 1881, the treasurer of the Cambridge Philosophical Society turned to the British Postmaster General with a somewhat odd request. Once a week, its treasurer, Dr. Pearson, hitherto showed up in the lobby of the local post office at a set time, perking his ears to "hear" the time announced by the telegraphist on duty at the London central telegraph office. Up until then, nobody took objection to Pearson's way of procuring accurate time for himself and his private collection of astronomical instruments. Now, however, Pearson was no longer alone in this practice, and the number of people gathering to hear the signal called had become an inconvenience to the post office staff. Pearson, as everybody else, was sent away and found himself bereft of his weekly audience with time.[68] Dr. Pearson of Cambridge was not the only citizen seeking to receive accurate time without having to subscribe to a costly and sometimes erratic service. Horace Darwin, another son of the eminent biologist Charles Darwin, suggested that a simple needle galvanometer be placed outside the Post Office "in some conspicuous position" and that the 10:00 a.m. time signal be made to pass through it, thus visually announcing the arrival of accurate time.[69]

Another unusual time distribution service developed from a makeshift arrangement to a flourishing family business. In 1835, then Astronomer Royal John Pond assigned his young warden Henry Belville the task of carrying an accurate chronometer around London in order to distribute the correct time to clockmakers and watchmakers. When Henry Belville died, his widow and later their daughter took over what had now become a successful business. Every Monday, Mrs. Maria Belville made her way to the Greenwich Observatory to have her chronometer checked and, certificate of accuracy in hand, set out for London to sell time. Ruth Belville, the daughter who took over the business, still had forty paying customers in 1908 and reportedly continued to carry out her daily trips until the 1940s. The shaky nature of even the most advanced time distribution technologies permitted improvised techniques of "hearing" the time and the business of the Belville family to thrive.[70]

The slew of new time distribution techniques of different origin and technological nature worked best in the confined environment of a factory complex or government institution. Once time distribution left the sheltered space of buildings and compounds, once it had to traverse more significant distances, it was subjected to the inclemencies of rain, ice, and wind, of falling branches or trees that severed wires and cut off electricity. Technology failed frequently, and accurate time remained hard to come by. In some more remote communities, no immediate need for accurate time was felt at all; timekeeping was casual. William Christie, Astronomer Royal at the Greenwich Observatory, volunteered his own experience in 1888, stating that "in districts which are not within the influence of railways the clocks of neighboring villages commonly differ by half-an-hour or more. The degree of exactitude in the measurement of local time in such cases may be inferred from the circumstances that a minute-hand is usually considered unnecessary. I have also found that in rural districts on the Continent [*sic*] arbitrary alterations of half an hour fast or slow are accepted not only without protest but with absolute indifference."[71] Moreover, public clocks erected in open air settings were for the most part fairly heavy and, by the end of the century, enormous tower or building clocks; the maintenance of these apparatuses posed serious challenges even to skilled personnel. As a consequence, the electric distribution of accurate time remained a flawed endeavor until at least the first decades of the twentieth century.

London, capital of the British Empire, financial center of the world, was held to particularly high standards in this world of capricious, multiple times. Living up to these expectations was not always easy. An increasingly time-aware and time-interested public viewed the city as deficient and negligent when it came to distributing and managing accurate public time. In 1906, the London City Corporation invited owners of public clocks to submit to a synchronization scheme for the improvement of the variant times on display in the city but received only one response. In a report of the same year, London's official city engineer arrived at the conclusion that of ninety-one public clocks only twenty-nine were synchronized. The report declared it highly desirable for the London City Corporation (the municipal governing body of the city) to mandate all owners of clocks overhanging public ways to keep them synchronized, as "at present it is a rarity to find any two alike." Two years later, the British Science Guild established a committee charged with the task of establishing the best possible way of synchronizing London's clocks. Not just synchronization was found to be wanting; the nationwide time signal service, too, frequently reported problems: new apparatuses did not mesh with equipment in place and time signals kept failing, prompting operators to transmit time verbally.[72]

The problems of spotty time signals and synchronization came to a head in 1908 in a controversy over mendacious timekeepers. In the winter of that year, a reader of the London *Times* unleashed an outpour of anger and contempt for the fits and antics of public time. Under the headline "Lying Clocks," Sir John Cockburn, vice chairman and founding member of the British Science Guild, launched a broadside against what he perceived to be the scandalous neglect by the government to uphold an aspect of public order. As "highly desirable as individualism is in many respects, it is out of place in horology," he wrote in a letter to the editors. Inaccurate public clocks ought to be forbidden by law and penalized with the removal of the "offending dial," he demanded. "A lying time-keeper is an abomination, and should not be tolerated."[73] The *Times* ran an article on the following day in response to Cockburn's letter and acknowledged that British time was in a deplorable state. But instead of penalizing private citizens for displaying if ever so slightly inaccurate time, the paper advocated getting municipal authorities involved. The *Times* responded that "it is, indeed, no very easy thing to ascertain the true time in many parts of this country. Railways are supposed to keep it but the clocks at roadside stations are of no very extreme accuracy, and even if they are frequently corrected they are very apt to be jarred out of truth by the shock of passing trains." The real complaint, the *Times* argued, was that London was still without one of the widely available systems for electrical clock coordination.[74]

Cockburn's letter to the *Times,* and the newspaper's riposte, triggered a series of writings to the editor that bespoke a widespread sense of irritation about the state of accurate time in the city. Clockmakers rushed to defend their profession's reputation, describing how complicated the mechanics of synchronizing two clocks were if said clocks were not of the same making. The Standard Time Company absolved itself from any responsibility for malfunctioning clocks by stating that contrary to public perception, it was not in charge of maintaining London's clocks. E. J. D. Newitt, the company's secretary, instead chastised the government for its "apathy" and lack of support for his business of synchronizing time. "Unlike Paris, Berlin, and other Continental [*sic*] capitals, London had no official recognized standard clocks." Rather, "in the present state of affairs every man's time is his own." One possibly French reader with the name of H. Berthoud picked up on the comparative European angle from which London looked particularly bad. "What is really needed is a good system of large electric clocks distributed in the principal thoroughfares of the metropolis, such as we see in most capitals of Europe. All along the Paris boulevards and in the principal streets of the French capital we see them, and it is the same in Brussels and also, I believe, in Vienna and Berlin. Why cannot London do the same?"[75]

Another reader emphasized the material losses accruing from mendacious timekeepers. "Here we are in London AD 1908, where time is said to be money, fatuously and impotently pottering about with innumerable 'lying clocks,' which are not only a scandal and a disgrace, but which inflict heavy pecuniary losses on the community." Later in the same year the "lying clocks" controversy led city authorities to investigate the supply of accurate time for government-owned clocks and watches owned by private individuals but displayed in public, but to no avail, as no agreement was reached. John Cockburn later chaired a committee at the British Science Guild that pursued a similar inquiry. Three years after the controversy the committee reached the sobering conclusion that with the exception of the Post Office, nobody had taken action to improve the misery that was London's public clocks.[76] In addition to the complications of technology and the absence of support from municipal government authorities, time distribution and synchronization faced resistance from traditional clockmakers. In a lecture before London's United Wards Club, John Winne of the Standard Time Company made a (surely self-serving) "Plea for Uniformity" and chastised traditional clockmaker associations. The profession of clockmakers and watchmakers had passed resolutions against electromechanical regulation, arguing that old-fashioned winders who performed the task manually once a week could do a better job. In this vein, the well-known clockmakers E. J. Dent and Co. had once called synchronizing devices "fakements."[77] The proliferation of different timekeeping synchronization technologies in the second half of the nineteenth century resulted first and foremost in a further multiplication and visualization of existing temporal variety. A swelling number of clocks all at sixes and sevens aroused the curiosity of the general public and fueled the general interest in the nature and meaning of time since the closing decades of the nineteenth century.

The litany of complaints about inaccuracy, paired with stiff opposition to innovations threatening to crowd out existing businesses, should give pause to unrealistic expectations about the thoroughness of supplying accurate, synchronized time until the early twentieth century. Creating legal mean times as an act of administrative politics was a mere first step that never ensured social compliance and always required follow-up adjustments and attunements. Even the law that set forth the use of Greenwich time in Britain was much more ambivalent than it is often portrayed in existing scholarship. In 1880, the British parliament passed the Definition of Time Act, making Greenwich Mean Time the legal time in Britain (but not Ireland). The Act stated that "whenever any expression of time occurs in any Acts of Parliament, deed, or other legal instrument, the time referred shall, unless it is otherwise specifically stated, be held in the case of Great Britain to be Greenwich mean time, and in the case of Ireland, Dublin mean time."[78] The main act of law by which a mean time was supposedly adopted for

Britain did not aim to ensure the nationwide observance of official Greenwich time but was instead intended to resolve confusion arising over the wording of legal texts. All in all, in 1924, a reader could therefore still express disbelief and marvel at the same time when writing that it was "no wonder our visitors from overseas comment upon the lack of official and reliable public clocks." "Of civilized countries," he mused, Britain was near alone in its neglectful attitude toward supplying accurate municipal time services. Only with the proliferation of time distribution through the radio and services such as the BBC in the 1920s and 1930s did accurate time really disseminate more widely into society.[79]

• • • •

TIME WAS ALWAYS MORE than a mere other object or convention that was standardized and internationalized in a worldwide process in the late nineteenth century, not least due to its foundational qualities as one of the two dimensions of human existence. Passing a law on time was one thing; ensuring that trains and other aspects of transportation and public life were scheduled in observance of said law was another. Supplying accurate and stable times was even more difficult. The production of abstract, uniform time was both institutionally and mentally a drawn-out and steep challenge. Transforming people's imagination and understanding of time was a change of a different order of magnitude. Historians have for long been beholden to the powerful narrative associated with E. P. Thompson.[80] Thompson's thesis about the move away from "task time" to "time-sense" and a time-conscious public was so compelling that subsequent generations have elevated it to axiomatic status mostly without further probing his arguments. Yet over the past decades, a curious gap between new research and the perception of Thompson's findings has opened. Scholarship has suggested the need to revise most aspects of Thompson's argument, but these challenges to his thesis have gone somewhat unnoticed.[81] In the perception of many historians, the standardization of time is still associated with the Thompsonian concept of time discipline, as a continuation and further abstraction of the process he ascribed to the eighteenth century. And yet, more than a hundred years later, as discussions about daylight saving in Europe demonstrate, even decades into the twentieth century, the internalization of an abstract, homogenous time detached from the rhythms of nature was far from complete, even among an educated audience of government officials and publicists. Mentalities turned out to be much more resistant than Thompson and many others envisaged.

The regularity of work rhythms and the temporal supervision of workers in factories and firms did intensify in the late nineteenth century. Work and work attendance were now more closely scrutinized and supervised than ever before, although "irregularities," such as Monday absenteeism induced by heavy drinking on Sunday, did not entirely disappear.[82] Some of the same companies that made time synchronization systems for public spaces supplied private businesses with the same apparatuses for installation at the workplace. Private time distribution services also offered a growing number of time clocks or punch clocks allowing employers to monitor their employees' working hours and punctuality ever more stringently. In factories, the question of "who watches the watchmen" was partially answered using new devices that monitored the overseer's tours of the workplace and his surveillance of workers and buildings.[83] More generally, a host of institutions contributed to the spread and internalization of discipline, temporal and other. Schools, hospitals, military barracks, and prisons formed a finely dispersed system of disciplinary techniques targeting the temporal routines and other habits of nineteenth-century citizens. These and other "complete and austere institutions" extended the life and the "regularity of the monastic communities to which they were often attached," as Michel Foucault has contended.[84]

Yet judging from the state of time distribution in early twentieth-century Britain and the difficulties of imagining an abstract time severed from nature's rhythms, if time discipline was a reality at this point it must have been able to burgeon without depending on or engendering a time consciousness that was rooted in a widespread presence of accurate mechanical clocks and abstract, homogenous time. It is helpful to recall that scholarship has found even early modern societies to have been highly time-oriented and time-disciplined. Research on time orientation and Japanese agricultural work in the seventeenth and eighteenth century has demonstrated that discipline and the valuation of time were by no means the prerogative of European industrial workshop and factory labor but characterized early modern agrarian labor as well. Capitalism was not the only, perhaps not even the primary, source of time discipline.[85] In early twentieth-century Europe, the time discipline of workers and employees had been honed, and clocks and watches were increasingly part of everyday life. Still, abstract time—a time untied from nature's (to some, still God's) rhythms—was near impossible to understand for many.

The connection that E. P. Thompson assumed between the availability of clocks and watches, factory work, and the internalization of time orientation is problematic. While public and private clocks and watches may have proliferated even beyond the sphere of luxury goods in the nineteenth century, with the exception of clocks overhanging the factory floor it was perhaps more the symbolic quality of timekeepers as status symbols and markers of modernity and progress that rendered them popular and bequeathed time with authority—rather than a desire for spreading or following accurate time and punctuality at the workplace and beyond. A demand for and interest in clocks and watches was not necessarily tied to industrial capitalism. In turn, the practice of acquiring accurate time by checking a personal timekeeper against the display or announcement of public time may have arisen out of a general fascination with time or, as in some bigger, early modern European cities, the complexities of urban life more than from internalized time discipline and adopted temporal norms.[86] Perhaps it was precisely this preoccupation with the symbolic qualities of time and its widening though less imposed than invited "rule" over so many aspects of human life that inspired a young French anarchist in 1894 to attempt to plant a bomb that would blow up Greenwich Observatory, as immortalized in Joseph Conrad's novel *The Secret Agent.*[87]

It requires reiterating that at different times, in different places, a glance at the clock tower and the setting of a personal watch bore different historical meanings and functions that were subject to change over time. In other words, supposedly "traditional" understandings of time as steered by inalterable natural rhythms could coexist perfectly with the presumably "modern" mechanical, abstract time of clocks and watches and uniform mean times, and with a growing submission to time. Accordingly, around 1900, the Thompsonian association of natural rhythms with timelessness and mechanical clocks with precious, abstract time would not have made sense to contemporaries. Late nineteenth- and early twentieth-century Europeans could abide by strict temporal disciplines without conceiving of time as abstract and homogenous, all the while eagerly setting their pocket watches to the authoritative time of one of the several nearby public clocks. Time regimes were manifold and coexistent. As recent research suggests, E. P. Thompson's proposed change in time orientation likely came before the onset of the Industrial Revolution, whether in Japan or Europe. Time discipline and tighter work rhythms, once adopted throughout the nineteenth century, did not rely on or cause a move away from task time to an understanding of abstract time. This disconnect between time discipline and notions of abstract time explains in part why economic demands and arguments for the adoption of mean times were so scarce when compared to rationalizing bureaucratic and cultural reasoning. Only in rare instances did business interests seek to exert pressure on governments with the goal of expediting the introduction of legal mean times. There was no need. Without abandoning alternative timekeepers such as the sun and biophysical rhythms, workers abided by the abstract time of mechanical clocks. At the turn of the twentieth century, then, industrial capitalism fetched unprecedented fortunes in the absence of a widespread understanding and application of accurate, abstract time. To paraphrase Timothy Mitchell, capitalism survived parasitically and flexibly, in this case amid a multitude of times. Anything else would be, to again use Mitchell's phrase, to ascribe to capitalism "a logic, energy, and coherence it did not have."[88]

Reformers discussing the social advantages and disadvantages of daylight saving targeted domestic socioeconomic groups as potentially prone to wasting their time outside work unless nudged into better behavior by a government that set clocks forward and back again each year. But such discussions about time management and useful activities did not go unnoticed in other parts of the world. Reformist-minded elites in non-Western societies keenly read Euro-American newspapers and magazines and absorbed the latest fashions and obsessions. Often, the presence of missionaries among their own societies added voices that called for the observance of time discipline, punctuality, and the careful division of the day into segments of time devoted to specific activities. As a consequence, non-Western reformers began to identify those guilty of time profligacy among themselves, in the ranks of their own societies. A global perspective on the reform of time proposes that their exhortations to fellow Arabs, Ottomans, and others to use time wisely instead of wasting it should be read in relation to Euro-American discussions about daylight saving.

CHAPTER THREE

# From National to Uniform Time around the Globe

In a number of countries in Europe and North America, mean times of even difference to Greenwich and distribution systems for disseminating that time had been instated beginning in the 1880s and 1890s. Yet by the first decades of the twentieth century, as evidenced in discussions about summer time, the adoption of stable, uniform time was far from complete in Europe and North America. Both technically and logistically as well as mentally and intellectually, the application of uniform mean time and summer time was hardly comprehensive. How did uniform time spread to eventually form a global grid of mostly hour-wide mean times? In the non-Western world, the move toward mean times and summer times was even further delayed. In Europe, the politics of adopting official mean times was a matter of bureaucracy and law, but the colonial state's interests in uniform mean times differed from that of Western European nation-states. Countries such as Germany, France, and Britain had labored long and hard to render time uniform as part of an uneven process of state-building and the constitution of clearly delineated territorial states. In the colonial world, where the presence of the state was often thinly spread, administrations initially had neither the will nor the capacity to effectively coordinate timekeeping. Colonial states thus began introducing mean times only in the early twentieth century, and in many locations, as late as the 1920s and 1930s. There was no system of Greenwich-based mean times prior to the middle decades of the twentieth century.

Some historians have suggested otherwise. It is commonly thought that a cast of colonial administrators eagerly discovered mean times in a quest to tighten the grip of the colonial state over its native subjects by improving administration and communication overseas as well as with the metropole.

In this vein, it has been proposed that the worldwide adoption of Greenwich time "not only greatly increased the efficiency of international business but enabled administrative services to be co-ordinated too, thus assisting the imperial state to integrate its overseas subsidiaries more effectively."[1] The "official deployment of GMT in 1884" is said to have heralded a new era of global timekeeping in allowing "people, towns, cities, ports, railways, and colonies to connect with metropolitan centers, and between each other, through a single space-time matrix." Such understandings are in part indebted to studies on telegraphy and railroads, technologies that have been shown to sometimes function precisely as such "tools of empire," seminal in undergirding or even enabling the expansion of European rule across the globe while subsequently sustaining it.[2] Yet the realities of colonial rule often looked more precarious. In the vast swaths of territory beyond Europe and North America, across oceans, deserts and steppes, and impenetrable jungles, sweeping narratives of successful time unification, discipline, and coordination quickly go amiss. As convincing as such conjectures about controlling colonial subjects are on the surface, they do not hold up to the scrutiny of archival research.

Seldom did the spread of mean times around the world originate in imperial centers bent on subjugating the periphery. Much rather, some of the earlier initiatives to introduce institutionalized countrywide times outside of Europe and North America were kindled by scientific associations, further underscoring the importance of such actors in transmitting ideas. For most, the movement of ideas about uniform time and summer time around the globe occurred in waves that followed highly regionalized patterns instead of determined decision-making processes in the capitals of Europe and North America. The news of introducing a mean time in one country or colony first attracted the attention of authorities in another neighboring locality and eventually induced administrators to take the same step and choose a mean time as well. Officials at home in Britain or Germany were mostly left out of the loop on these decisions. Such regionalized solutions for time unification often initially disregarded the concept of hour-wide time zones based on Greenwich. Many countries and colonial administrations around the world introduced mean times in the 1880s and 1890s, following the efforts of scientific internationalists and mirroring European countries in the same period. But with only rare exceptions, these were the times of national capitals or important colonial cities, mean times that stood in no uniform relation to the prime meridian at Greenwich.[3] Mean times, moreover, were adopted late and changed frequently, adding further instability and multiplicity of times. Time reform was certainly contagious, as news about European mean times traveled to other parts of the world.

But it spread as national, not universal, homogenous time. The introduction of countrywide mean times outside Europe and North America therefore corroborates what European time politics have already revealed: the mutually constitutive relationship between state-building and regionalization on the one hand and interconnectedness on the other.

• • • •

THE FIRST WAVE of efforts to introduce territory-wide mean times outside of Europe and North America occurred roughly between 1897 and 1906. It was precipitated neither by European colonial governments, nor by one of the colonial ministries in London, Berlin, and Paris in charge of administrating overseas possessions. Instead, it was sparked by two memoranda drafted in 1897 and 1898 by the Seismological Investigation Committee of the British Association for the Advancement of Science and the Royal Scottish Geographical Society, respectively.[4] Such associations and organizations had already played an important role in the nationalization of time within single countries like France and Germany. They flourished as part of a trend to popularize scientific knowledge in the second half of the nineteenth and early twentieth century.[5]

One of these documents, the correspondence from the Royal Scottish Geographical Society, had prompted the initial inquiry by the General Post Office into possibly extending Greenwich time to Ireland. Scientific associations systematically inundated ministries and other government institutions with cheaply printed circulars and pamphlets. Both communications were addressed to the British Colonial Office as well. The correspondence urged improvement in the current state of mean time application in the British Empire, declaring the absence of uniform mean times unworthy of an imperial power of Britain's stature. It was enough to shame Joseph Chamberlain, the secretary of state for colonies, into opening a review of the status of time unification across the empire. In 1897, Chamberlain mailed off a circular to all British possessions asking for information on the time currently kept within the territory and the difference between the time in use and Greenwich time. The colonial secretary did not move to force the hands of local colonial governors and other administrators to adopt mean times; however, his inquiry inspired others to take steps. Actions percolated slowly through administrative chains of command, but around 1900, spurred by London's probe, several colonial governments suddenly took an interest in mean times. Laws and resolutions implementing time changes not only took several additional years to come to pass but also evolved without much oversight and intervention from the metropolitan center.[6]

Among those possessions that acquired a time zone early was the British Federation of Malay States, an assemblage of four protected states in the Malay Peninsula. In a pattern of decision making that betrayed the absence of a directed movement and push for coordination, it was announced in the official gazette that the federacy in 1900 had adopted the local time of Kuala Lumpur as the mean time for the entire Federation of Malay States. The note announcing the change in the official gazette set off an exchange among officials: which time to choose in the first place, how to best coordinate with other British imperial possessions in the region, and what relationship the new mean time should have to Greenwich time. Different proposals were now put forward, and the next question raised was whether to include the adjacent colony of Singapore in the time zone scheme.[7] Singapore officials were hesitant. Time especially for the captains of shipping vessels was regulated by a time ball signal given every day at UTC + 6:55:25, local Singapore time. "The adoption of a local time in conflict with this would cause great confusion," and resulting mistakes among shipmasters would be "disastrous," Singapore officials warned. Singapore had the upper hand in the end and convinced officials in the Malay Federation to adopt Singapore local time as of January 1901. But the arrangement lasted only four years until in 1905 it was found more convenient to switch to the time of the 105th longitudinal arch. That time, UTC + 7, was used in the British Straits Settlements as well as French Indochina and Siam. The time seven hours ahead of Greenwich had become something of a regional time, integrating colonial possessions even across the borders of different imperial powers.[8]

A slew of other locations in South, Southeast, and East Asia chose mean times around the same moment. In 1903, the British colonial administration adopted the time of UTC + 8 as the official time for Hong Kong, a time zone that soon welcomed other cities on the Chinese coast; Shanghai, however, continued to follow local time, which ran six minutes behind the new regional time zone. British North Borneo adopted UTC + 8 in 1905, primarily because important regional centers like Hong Kong, the German colony of Kiaochow, the Philippines, and Formosa had adopted the same time. Regional considerations eclipsed both international and imperial motives in choosing times.[9]

• • • •

CHAMBERLAIN'S CIRCULAR about differences between local and Greenwich time sparked a similar move toward regionally uniform time in Southern Africa. Already in 1892, the Cape Colony had made the time of 1.5 hours ahead of Greenwich its mean time. On that occasion, a time ball service was inaugurated in a handful of cities, receiving time signals from the Royal Observatory at Cape Town.[10] At the same time in the early 1890s, the British colony of Natal, separated from the Cape Colony only by a small area of associated treaty states and protectorates, adopted UTC+2 as its official time. Like so many cities and states in these years, Natal was struggling to keep accurate time. The central clock at the Natal Observatory was prone to breaking down, and delays in the negotiations with the manufacturer held up a replacement that had been ordered as early as 1889. Three years later, when the new official mean time was to be introduced, Natal was still in "urgent want" of a new apparatus. Perhaps in part because the Natal timekeeper proved unreliable, and more certainly because keeping two different times within considerable proximity was cumbersome, South Africa's multiple times were fused into one in 1902. That year, the time two hours ahead of Greenwich was adopted as the legal mean time for railways, telegraphs, and other public purposes in the Cape Colony, Natal, Transvaal, the Orange River Colony, Rhodesia, and, remarkably, Portuguese East Africa.[11]

The British governor of the Cape Colony had initiated a concerted effort among regional colonial governments to adopt a common mean time and in this endeavor had reached out to the Portuguese. As part of the same initiative, British authorities also contacted the neighboring German colonial administration in the colony of Southwest Africa. Unbeknownst to the Cape Colony governor, his plan to unify the several times in use throughout the region spurred the first German administrative encounter with the times observed in overseas possessions. The impetus for time reform did not originate in Berlin but surfaced regionally, in this case with a foreign colonial power. German Southwest Africa was the first German colony to obtain a legal mean time. Even before British authorities turned to their German counterparts, the German colonial government in Windhoek was already receiving daily time signals from Cape Town's observatory. The time of the Cape was deployed to carry out comparisons among different clocks and instruments in the possession of the German colonial government.[12]

Despite these existing ties of time distribution, German colonial officials immediately suspected political motives behind the British time unification plan. Regional integration was too closely associated with spaces of hegemony and spheres of influence to pass as innocuous. Similar to the way in which Germany's time zone delineated Central European space, German administrators now associated Britain's move with her presumed hegemonic aspirations in Africa. "This has probably happened with regard to England's plans for a Greater Africa ('grossafrikanische Pläne') and her drawing close to the time of UTC+2 that is already the predominant time in Egypt,"

Germany's Foreign Office remarked. While Germans harbored suspicions, the reason behind ultimately not participating in the regional time scheme was a different one. When the British governor approached the German administration in this joint endeavor between the British administration, the Germans, and the Portuguese in Mozambique, German colonial officials declared UTC+2 to be unfavorable to local conditions: solar time diverged too widely from the proposed time. In an age without air conditioning and other amenities of modern life, the extremes of the climate in colonial situations dictated that life remain finely calibrated to the ways of the sun.[13]

Piqued by the British requests, German colonial officials in Southern Africa warmed to the idea of time unification generally. Their time of choice was now Germany's own Central European Time, one hour in advance of Greenwich. As of November 1903, German Southwest Africa thus adopted Central European Time as its legal mean time. As in Europe, time distribution and the supply of stable, accurate time was a persistent problem in Southwest Africa. Laments about the state of timekeeping in the colony never ceased. Germany's colonial time woes, persisting for more than a decade, are a telltale demonstration of the confusion and informality that ruled colonial timekeeping during these decades. In the specific case of German Southwest Africa, several years after settling on a new mean time for the colony, the question of what time was in fact observed in the territory remained foggy and attempts to unify the protectorate's times had dissipated.[14]

In 1908, the Colonial Office in Berlin received a letter from a European citizen residing in German Southwest Africa with a complaint about inaccurate information in a guidebook on life in the German colony. The guide indicated that the official time in Southwest Africa was UTC+1, while he had found local Windhoek time to be setting the pace in the colony, the time of the administrative capital. The Colonial Office in Berlin set out to assert its authority and chastised the German governor in Windhoek for failing to comply with previous orders. After all, Central European Time had been officially adopted in 1903, as the Colonial Office reminded Windhoek. Much to Berlin's surprise, however, this turned out to be false information. The Colonial Office soon found out that the denizens of the protectorate had long since taken matters of time into their own hands.[15]

In the earlier years of German rule after the colony had been set up in 1884, the German administration received daily telegraph time signals from Cape Town, sent to the coastal town of Swakopmund. Here, time was converted to local Swakopmund time, which served as the basis for calculating local times throughout the protectorate. Yet cable transmission had shown itself to be unreliable; clocks at Swakopmund were plagued by frequent breakdowns; and public servants charged with transmitting times lacked the skills required for performing these tasks. Daily aberrations of the time signals therefore frequently amounted to anything between six and fifteen minutes. It was to ameliorate this imbroglio of timekeeping that Windhoek time had been made the official mean time for German Southwest Africa in 1902, as the Berlin Colonial Office now learned. To eschew the inaccuracies that necessarily accrued when time was transmitted from Cape Town and subsequently converted, time was now derived from a more precise naval chronometer, newly installed at the land survey office in Windhoek itself.[16]

The Colonial Office in Berlin could hardly conceal its dismay in the face of these revelations. The time legally in force since 1903 had simply been ignored. And more discoveries were in store. Berlin was informed that absent any enforced colony-wide regulation, the different branches of the colonial administration as well as railroads and postal services were using certain locally derived times. Bernhard Dernburg, the colonial minister himself, therefore ordered a time (Central European Time) that had never been introduced in the first place as of June 1909; uncoordinated makeshift arrangements created nothing but a "completely untenable situation," he grumbled. Another three years passed until in 1912, the Colonial Office took steps to investigate what had become of its directive. In response to this probe, the District Office in Swakopmund revealed that it was setting its clocks and watches after the time observed on the Otavi railway, a line said to be in possession of a reliable chronometer for the daily control and adjustment of the railway's clocks. In response to the 1912 inquiry, the district office at Lüderitzbucht simply relayed that in 1905 or 1906, one was not really sure when exactly the governor had designated railway time to be the protectorate's official time, and that the order had since been lost.[17]

A similar situation took hold in Togo, often admiringly called Germany's "model colony" (Musterkolonie). In colonial settings, debating a change of mean time was often a discussion about the rhythm dictated by an unforgiving and scathing sun. Several colonies therefore followed unofficial summer time schemes by adjusting office hours during the summer months even before actual summer time orders were passed. In colonial contexts, daylight saving was primarily heat avoidance. When Togo's official gazette in 1914 announced that local time would be advanced fifteen minutes each year between November and April to accommodate the seasonal variations of daylight and heat, the question of legal time in Togo for the first time caught the eye of colonial administrators in Berlin. As the governor of Togo explained, the announcement was based on an arrangement dating back to 1909 (of which Berlin obviously had not been informed), in which the colonial government had enlisted missionaries and the Postal and Telegraphic administration to observe a quasi summer time "in order to eliminate the influence of the early sunset time on working hours during the months of October through December." It was Berlin's turn to scoff at yet more evidence of unilateral action. The Colonial Office suggested the most suitable time zone for Togo's location would be Greenwich time proper only to be informed that in 1907, the time of the town of Lome had been declared Togo's mean time.[18]

In German East Africa, it was an initiative by the railways that led to the adoption of a mean time. The East African Railway Company as the main operator of rail traffic in the colony contacted the colonial government with a plan to adopt a legal mean time for the protectorate as a whole. Previously, life in East Africa followed local times, announced by a gun that was fired daily in bigger cities like Dar es Salaam.[19] Missionary stations and plantations not connected to the telegraph network did not even have a time gun or telegraphic time signal to observe but gauged time by means of a simple sundial or by merely deriving the time with the help of an astronomical almanac, based on the position and height of the sun. East African railways initially used their own mean times. Once a week, the railway officers at the company's headquarters received a time signal from Dar es Salaam via the post office and set the main company clock in accordance with it. Differences of one or two minutes were commonly disregarded; under normal circumstances railway officials did not adjust watches for another week until a significant aberration had accrued. The company's main timekeeper held at the headquarters was thus turned into a coveted object that was to be safeguarded and secured, with the keys stowed away in a safe.[20]

Such practices and the widespread inaccuracies they entailed led the colonial administration to consider the introduction of a colony-wide mean time beginning in 1913. Skeptics argued that introducing any kind of mean time would involve unsustainable divergence from apparent solar time and predicted problems. Proponents of time reform prevailed. The colonial government in Dar es Salaam initially favored the meridian running through the town of Kilimatinde, as it lay at the near median point of east-west extension of the colony overall. Soon, the hope was to convince Britain to adopt a time conjointly with German possessions.[21] Once such expectations fell through, German officials in Dar es Salaam proposed the introduction of so-called Moshi time, the official time deployed in British East Africa since 1908. Lingering unease with adopting the time of a foreign colony among some administrators was brushed off as "parish-pump politics"

(Kirchtumpolitik). At the end, the time of UTC + 2:30, roughly the time of the longitudinal arch running through Moshi, a town perched on the lower slopes of Kilimanjaro on the border with present-day Kenya, was adopted as German East Africa's legal time in October 1913. German colonial officials thus prioritized regional practicalities all the while disregarding concerns about adopting a colonial competitor's time and complying with global uniformity alike.[22]

German colonial mean times introduced during the years 1912–1913 came at the same time as French, British, and Latin American efforts. The years prior to World War I therefore saw another spike in time unification. In the French empire, time unification occurred in conjunction with the switch to Greenwich time in metropolitan France. Most French territories appear to have adopted hour-wide zone times in 1911 and 1912. British possessions in the Caribbean—the British West Indies and British Guiana—followed UTC − 4 as of 1911. Officials in London questioned the utility of adopting a mean time in the first place, given the "lack of . . . exactitude in the best chronometers in the West Indian climate." In Latin America, Chile adopted UTC − 5 in 1910.[23]

• • • •

FOLLOWING WORLD WAR I, another wave of time standardization rolled in, this time in Africa and Latin America. The war overturned politics in Europe by precipitating the breakup of the old landed empires of the Habsburg, Russian, and Ottoman kind. Newly carved out nation-states, or as in the case of the Soviet Union, an unprecedented experiment, emerged in their stead. In the colonial world, the defeated central powers Germany and the Ottoman Empire were stripped of their colonial and imperial possessions. Instead of achieving independence, these territories merely changed overlords, however, to be placed under "Mandate" power at the League of Nations, a form of tutelage administered primarily by Britain and France. Regime changes and postwar reordering brought with them administrative reorganizations and, often, time changes. One of these time changes, in Russia and the Soviet Union, would merit a comprehensive investigation in its own right, for longitudinal extension, ethnic and religious diversity, and the magnitude of political change after the Russian Revolution in 1917 make for an intriguing case: Russia and eventually the Soviet Union gradually carved up its vast territory into eleven time zones between 1919 and 1924 after following St. Petersburg time on all railways since the 1860s but otherwise observing local time.[24]

In Africa, many colonies had instated mean times in the 1890s by choosing the time of a major city to be deployed throughout the territory. Now, after

World War I, times were changed to reflect hour-wide time zones. In West Africa, Nigeria (UTC+1), the Gold Coast, Ashanti, and Northern Territories (UTC) adopted new times in 1919, and the Gambia (UTC−1) followed in 1933. To the east of the African continent, Tanganyika Territory, the British-administered League of Nations Mandate that had previously been German East Africa, switched its standard time from UTC+2:30 to UTC+3. Kenya too changed its time to run three hours ahead of Greenwich in 1928, only to move back to the old UTC+2:30 in 1930.

During the 1930s, the frequent to and fro and the resulting regional time differences gave cause for concern among Europeans residing in eastern Africa. In 1936, the Associated Chambers of Commerce in eastern Africa spoke out against the regional patchwork of time standards. It was inimical to commercial interests and created unnecessary confusion especially in transportation, it was held. Kenya ran on UTC+2:30, Tanganyika on UTC+3, and bureaucrats in both places were reluctant to budge an inch. A meeting of east African colonial governors eventually produced a proposal to use the time 2:45 minutes ahead of Greenwich. For a brief moment, Kenyan colonial administrators worried to be in violation of accepted rules and conventions. It was the Royal Astronomer at Greenwich himself who replied that while it was customary to adopt mean times differing from Greenwich by an exact number of hours or half hours, it was not contrary to "any decision reached as a result of the Washington Time Congress." Moreover, the Astronomer Royal proceeded to adduce several other British dependencies, all of which were in observance of uneven mean times: the Federated Malay States and the Straits Settlements (except Labuan and Christmas Islands) changed the time initially implemented and since 1930 followed UTC+7:20; the Maldives observed UTC+4:54; Aden and British Somaliland UTC+2:59:54; Labrador and Newfoundland UTC−3:31; British Guiana UTC−3:45. Kenyan officials needed no further convincing and made the transition to UTC+2:45 as of 1937.[25]

In Latin America, debates in the scientific community over the value of mean times had been stirred as early as time unification was discussed at pertinent international conferences held in Europe in the 1880s, but action did not follow until the first decades of the twentieth century. As expressed in the burgeoning landscape of scientific associations and journals, countries like Brazil, Argentina, and Chile treated accurate and uniform mean times as one aspect of scientific modernity to which the new middle classes in these societies aspired. Latin American scientists actively followed the work of their peers in European and American outlets and, if feasible, at international conferences. Scientific modernity was embraced as a proof of status among the so-called advanced, civilized nations. Whether in urban planning, medicine and disease, or identification and documentation, societies in Latin America were not immune to the global infatuation with science and rationality as tokens of modernity, so characteristic of the global fin-de-siècle around the world.[26]

Following these early conversations about mean times in Latin America, a concerted regional push emerged from regional scientific internationalism. In 1901, a section of the second Latin American Scientific Congress in Montevideo passed a resolution that supported the adoption of time zones in Latin America. A following congress held in Santiago de Chile in 1908–1909 formulated a similar proposal. Delegates were now instructed to work toward convincing their governments back home to take action. Two years later, the Latin American Scientific Congress held its meeting in conjunction with the International Conference of American States, which in 1910 changed its name to Pan-American Union. When the Conference of American States first met in the 1880s and 1890s, the Pan-American movement was viewed by the United States as a tool to consolidate America's hemispheric power in the tradition of the Monroe Doctrine by influencing decisions in Latin American countries. At the twin meeting of the scientific congress and the more politically oriented Pan-American Union in 1910, scientists declared that "it would be very useful for those countries which have not adopted until now the Greenwich time zone system to adopt it from January 1, 1911." Among the scientists who sponsored the proposal were Charles Dillon Perrine, an American who headed the Argentine National Observatory at Cordoba, Friedrich Wilhelm Ristenpart, a German leading the Chilean observatory in Santiago, and Richard Tucker, another American who helped run the San Luis observatory in Argentina, an institution that was funded by the Carnegie Foundation.[27]

Peru might have been the first country in Latin America to set its clocks in accordance with Greenwich time. It was an invitation to participate in an Italian conference in the 1890s that raised interest in uniform time among Peruvian scientists and politicians. Italian scientists had persuaded the Italian government into organizing a meeting that would make Jerusalem the universal meridian. Likely in an attempt to recruit Catholic countries, Peruvian authorities received an invitation.[28] While the Italian government had lent the project its seal of approval, it was the Academy of Sciences in Bologna where the Barnabite monk Cesario Tondini de Quarenghi was the actual agitator for Jerusalem. Federico Villareal, a professor at the College of Sciences in Lima, Peru, thought it unlikely that Jerusalem stood a chance, given the deliberations of the Washington meeting in 1884. But he acknowledged that the resolutions passed in Washington, DC, were by no means binding. Few countries had in fact begun to apply the time zone system. Peru, he opined, ought to be one of them in a time when the wires of the telegraph connected an ever-growing swath of Latin American territory. Moreover, for the advancement of agricultural and other sciences it was mandatory to conduct the most precise meteorological observations, all of which required an exact determination of time. Villareal and his Latin American contemporaries had internalized the same quest for precision and accuracy that drove their Euro-American colleagues.[29]

At the beginning of the twentieth century, the country still lacked a uniform time, and Peruvians showed an ongoing interest in the topic. In 1907, the Lima Geographical Society under the penmanship of its president Eulogio Delgado actively lobbied the Foreign Ministry to promote time unification in Peru. In major cities such as Lima, municipalities, post offices, railway stations, and other public clocks displayed "considerable" variation. The same "irregularity" plagued other cities. The government needed to step in to "safeguard people from being disoriented."[30] For choosing a time zone the scientists looked to the United States. It happened that the time of UTC − 5, which regulated life on the East Coast, ran roughly through the middle of Peruvian territory, making it a good fit for the Andean state. Following the initiative by the Lima Geographical Society, the time five hours slow of Greenwich was adopted in June 1908. Execution and implementation of the new time were not without glitches. In 1910 it was reported that the clock at the majestic cathedral of Cuzco in the southeast of Peru—"the only standard time that we possess to regulate public offices in El Cuzco," as one contemporary claimed—suffered damage and ceased to function for more than a week. Local citizens were at a loss for they mistrusted the other available alternative, the municipal clock, "which always disagrees with the one mentioned before." The author of an article describing Cuzco's time woes urged municipal authorities to take seriously the resolution passed by the government in 1908 and to enforce the application of uniform time throughout the country. Only with the help of more energetic time politics would reliable time be more abundant and confusion arising from a defect in the sole trusted clock prevented in the future.[31]

In Brazil, it was a confluence of government initiatives to harness the state as the engine of modernization and the emergence of a national scientific community that impelled the adoption of uniform time. Back at the 1884 prime meridian conference, the Brazilian delegate had been ordered by Emperor Pedro II to vote with France. Luis Cruls, director of Rio de Janeiro's Imperial Observatory, went even further when he voted against Greenwich as the prime meridian instead of abstaining as France did. Like the French delegate, Cruls explained his country's vote by underscoring the need for a neutral meridian instead of one that ran through sites of national significance. It may have been an ongoing sweltering conflict with Britain over slavery in Brazil that led Don Pedro to side with the French, or merely an attempt to curb British influence in the world and perhaps indirectly, its significant informal economic and financial might in Latin America. Moreover, French and Brazilian astronomers had maintained close ties in those years. Between 1874 and 1881, the observatory in Rio de Janeiro was run by the French astronomer Emmanuel Liais.[32]

Another thirty years passed until Brazil established an official mean time for the country. In 1911, a Brazilian navy captain gave a speech at the Brazilian Historical, Geographic, and Ethnographic Institute, an institution that closely resembled Euro-American scientific societies of the period. Radler de Aquino spoke of international communication and transportation and commercial and scientific needs when he argued for Brazilian time zones. And as usual, he detailed instances of extreme temporal disarray: in Santos, for instance, the main port of the state of São Paulo, railways used Rio time while for official government purposes São Paulo state time was observed; there was local Santos time as well. Other states similarly used their own time. One year later, the Brazilian astronomer Nuno Alves Duarte da Silva attended the Paris international time conference in 1912. He returned convinced that Brazil was in need of a time zone system and set his mind on persuading the Brazilian Congress. He succeeded, and after the new time law came into force in 1914, Brazil was divided into four time zones counted from the zero meridian at Greenwich. Other countries in Latin America followed after World War I. Uruguay, having originally set its clock to Montevideo time in 1908, switched to UTC−4 in 1920. Guatemala followed UTC−6 beginning in 1918. In Argentina, where Cordoba time had been used as the countrywide mean time since 1892, UTC−4 was followed as of 1920; Uruguay chose UTC−3:30 in 1920; Mexico used UTC−6 as of 1922.[33]

The 1920s and 1930s were a time in which small islands and lesser territories adopted mean times. In these particular locales, mean times most likely followed from a decision to extend time zones to apply on the open sea. Until 1920, oceans and seas remained timeless. Even in the abstract plans of scientists and bureaucrats who worried less about practical matters of application, the system of time zones never truly girded the globe. There may have been imaginary time zones between continents, allowing one to arrive at the correct time zone for the American East Coast when counting from Greenwich. But ships did not follow these zones in navigation, for officially they did not exist. Timekeeping at sea was uncharted, and improvised solutions prevailed. British ships, as one example, usually kept two times. One was the time of a chronometer taken aboard and set to Greenwich time, used to navigate the ship. Another set of timekeepers regulated life and record keeping on board. These clocks normally showed solar time adjusted at noon, and if a ship crossed significant distances between noon on one day and the next, adjustments were made over the course of the hours. Such alterations relied exclusively on the judgments of officers in observing the position of the sun, rendering the times kept on seaborne vessels incomparable.

In the past, it had proved difficult to identify the time of an event noted in a ship's records. During World War I, the incomparability of nautical positions caused particular problems. No end of fighting was in sight and the conflict was intensifying by drawing a growing number of countries into its vortex. In 1917, the French Navy began to examine the possibility of establishing time zones at sea, counted from the zero meridian at Greenwich just as landed zones had been determined. In May 1917, a conference of British, French, and Italian officials was held in London. It was the height of the Great War with the Americans now officially battling Germany and her allies. Germany had resumed unrestricted submarine warfare in January of the same year and was sinking American ships in the North Atlantic. Yet none of these developments sufficed to convince the group of military and scientific experts gathered in London to observe time zones at sea; the French scheme was deemed impractical. Three years later, another such meeting convened, again following a French initiative. The conference of 1920 now concluded that establishing time zones at sea outside of territorial waters was the most practical method of obtaining "uniformity in time-reckoning at sea." The French Navy drew up a scheme of boundaries for maritime time zones, and time changes aboard ships were henceforth to be made hourly in accordance with the zones crossed.[34]

With the seas now charted, several smaller island territories decided on mean times. Many of these localities followed times that varied from Greenwich by half- and quarter-hour differences. There was still no consensus that mean times had to be set in accordance with hour-wide time zones. It was only gradually that odd minutes and half-hour differences were being ironed out. In 1929, Bermuda passed a Time Zone Act that changed the colony's previous time of UTC−3:45 to UTC−4 as of 1930. Two years later, the Solomon Islands switched from UTC+10:40 to UTC+11; Gilbert and Ellice Islands in the western Pacific had previously followed local time and now used UTC+12.[35]

The gradual espousal of mean times in the interwar years was related to another aspect of time politics. Beyond Europe and North America, time unification raised the question of time differences between solar time and a new mean time in augmented ways. In the humidity of the tropics, under the drastic temperature changes in the deserts and steppes of sub-Saharan Africa, in the searing dry heat of the Middle East summer, colonial officials perceived themselves as living in a permanent state of exception, battling climates no less hostile than the indigenous populations they sought to conquer and rule. In every moment, the vicissitudes of the climate threatened to derail the colonial enterprise. Daylight was intimately linked to temperatures and humidity; settling on a mean time that was at ever so slight a variance with the solar rhythms of temperatures and humidity gave reason for grave concerns among colonizers on the spot.

The treacherous climates of the colonial world agonized administrators as early as mean times caught their attention. A widely held belief pervading medical discourses of the time was that Europeans were by "racial" disposition ill-equipped for physical work in tropical climates. Such views in turn fueled a determination among government officials and advocates of colonial rule to employ native labor. But for many administrative tasks the colonial overlords were reluctant to rely on untrained indigenous workers. German or other European office clerks therefore found themselves at the center of attention in colonial daylight saving debates.

When colonial officials began to discuss the selection of uniform mean times in German colonies in Africa, they shook their heads in disbelief at the uninformed suggestions made by the higher-ups in Berlin. As complete strangers to the local conditions on the ground, their proposals "completely ignore[d] the tropical circumstances of time," members of the government in Dar es Salaam accused staff at the Colonial Office in Berlin. In Dar es Salaam, Europeans could engage in recreational activities only as long as the sun was up, "before swarms of mosquitoes" and other carriers of infectious diseases ventured out.[36] Tied to the climate was the question of working hours. Colonial officials eagerly strove to impose some form of unity in the working hours for clerks in different branches of the administration. But opinions varied greatly as to what the most feasible solution would be—working hours with longer midday breaks, working hours that began early in the morning and ended early in the afternoon, or, as railway administrators demanded, office hours that extended into the late afternoon. Whatever the individual preferences, it was not uncommon for colonial administrations to observe informal summer time by starting and ending work earlier or later during the summer and winter. The judicial administration in German Southwest Africa ordered a summer time of sorts to be applied between October and April, when office hours would begin and end half an hour earlier than during the remainder of the year. In all of this, the lack of accurate clocks and time distribution was mentioned as an aggravating factor.[37]

By the time colonial administrations considered regular and official summer times for overseas possessions, German colonies had ceased to exist as such and had been handed over to new rulers. Other colonial powers such as Britain, whether in former German colonies and now League mandates or other spots of the empire, were keenly interested in applying summer times. Even more so than with mean times, summer times in colonial settings disposed colonial governments to take steps that increased the heterogeneity of times in use. In 1919, the Gold Coast (Ghana as of 1957) made Greenwich time its legal time and simultaneously legalized a summer time of UTC – 00:20 minutes from March to October.[38] Similarly, in 1932, the government of Sierra Leone wrote to the Colonial Office in London and announced its plan for daylight saving time. From October to March, the clock was to be advanced twenty minutes. Like their counterparts in German colonies earlier, British colonial officials in Sierra Leone deemed it crucial to "provide both Europeans and Africans with more time for daylight recreation which is so essential to health in the Tropics." The local chamber of commerce and the Freetown City Council had already approved of the plan. London, however, was not convinced. "I don't think any case is made out for such an important change," a member of the Colonial Office scrabbled onto the correspondence he received from West Africa.[39] What bugged the official was not the "irregular" nature of mean time that, for a considerable part of the year, was at variance with the time zone system, but the idea to pass such an important act for a mere gain of twenty minutes of daylight. Sierra Leone was eventually given a pass, but with the outbreak of war in 1939, as part of the Emergency Powers (Defense) Act, switched to Greenwich time year-round. In February 1946, the wartime amendment was made permanent.[40]

In Kenya, colonial summer time was a particularly contested affair. Kenya had adopted the time of UTC + 2:30 in 1920. Several times throughout the 1920s, Kenya's legislative body considered "daylight saving time," as participants in the debates and administrators alike referred to it in their exchanges. Yet what was being discussed was simply the adoption of a mean time with a more favorable relation to daylight by permanently changing the colony's time to UTC + 3 for the duration of the entire year.[41] In the understanding of contemporaries, colonial mean time was inseparable from the daylight effect, and consultations over several years therefore referred to the measure of changing the territory's mean time permanently as "summer time." The first attempt to improve the daylight situation in Kenya dated back to the early 1920s, when the legislative council discussed "a zone time, as standard, that will provide a better utilization of the daylight hours throughout the year."[42] In the following years, time found its way onto the council's agenda seven more times before a solution was reached in 1928. That year, one Captain Schwartze, a European resident of the East African colony, brought the issue of daylight saving time before the Kenya legislature after a previous attempt was struck down by a slim majority of three votes. Schwartze was hopeful that matters would now finally be resolved.[43]

The narrow defeat prompted the appointment of a committee charged with gathering evidence from a broad range of groups and professions on the merits and disadvantages of the proposal. The adoption of a mean time with daylight saving advantages was motivated by the same concern for health and recreation in hot, tropical climates that drove German administrators at the beginning of the century. If the clock were advanced by half an hour, the European population in the colony stood to gain half an hour between the end of work and the fall of darkness. The committee on daylight saving explicated, "Medical evidence also strongly supported this view" and stressed the fact that in the tropics the "ordinary individual requires plenty of exercise, and that recreation every night brought considerable benefits, in the form of increased vigor, health, and vitality."

As in Europe several years earlier, counterarguments preferred means other than legally mandating the advance of the clock for achieving the same results. Why not oblige shops by law to close at half past four instead of five o'clock and thus allow employees "to get their much-needed exercise," it was asked.[44] One council member, Lord Delamere, an influential British settler in Kenya, moreover struggled to grasp time as an abstract, movable grid. Delamere cautioned that under the new time regime, office clerks would have to begin work at 7:30 a.m. instead of 8:00 a.m. Delamere, like his contemporaries in Europe, imagined a time regime under which meal times, sleepiness, and other rhythms stayed fixed. People would not be able to eat earlier and go to bed earlier after rising an hour earlier in the morning. It was 1928, and Europeans still grappled with the characteristics of abstract time.[45]

Timekeeping remained in a state of improvisation even in Kenya, a settler colony with significant European presence. Due to a lack of accurate timekeepers and time distribution in the colony, many farms operated on sun time. One representative of the legislative council held no illusions in this regard. "All I can say from observation," he reported, "is that the average clock on the average farm is merely a mechanical contrivance for dividing the combined diurnal and nocturnal periods into twenty-four equal parts. Its relation to the actual time in general is anything from half to three-quarters of an hour inaccurate. . . . The relationship of these hours to the standard time of the country is very often negligible." In the end, the proposal to set Kenya's clocks forward by thirty minutes to assume a mean time of UTC + 3 passed with nineteen over twelve votes. The hard-fought victory did not preclude Kenya from switching back to UTC + 2:30 only two years later. Resistance proved too much to overcome just yet.[46]

More than a decade later during World War II, several British colonies, protectorates, and dominions were made to join the British war effort by implementing daylight saving times for the purpose of saving energy. In subsequent years, however, the experiment was dropped in several locations after protests among local populations. In South Africa, "the most bitter opponents were the farmers, who were unable to change their daily routine to fit in with 'summer time.' " Australia reported similar misgivings. "Last year there was much dissatisfaction among a large section of Australians, particularly country people, who said that daylight saving upset the routine of their lives, which had different needs from those of city dwellers," the *Times of Australia* stated.[47] After a one-time trial in 1942, the British Protectorate of Bechuanaland and the administration of Southwest Africa (the former German colony was now under British and League of Nations tutelage) decided not to reinstate summer time the following year. Europeans had rejected summer time, and in any case, "the African population are but little affected as they take their time from the sun." Territories in the Arab Middle East had introduced mean times of UTC + 2 with the advent of Mandate rule and the British and French presence in the Eastern Mediterranean in the early 1920s. During the war, the British Resident Minister in Cairo, acting "in the interest of uniformity," announced summer time in 1944 and 1945 for Egypt, Palestine (present-day Israel/Occupied Palestinian Territories), Transjordan (present-day Jordan), Syria, Lebanon, and Cyprus.[48]

• • • •

COLONIZED POPULATIONS are near absent from the correspondence on time changes gathered together in the archives of former colonial powers. Attempts to discipline colonized subjects with the help of elaborate timekeeping and time distribution schemes did not leave traces in the circulars mailed back and forth across the Mediterranean, the Indian Ocean, and the Atlantic between officials on site and colonial offices in Berlin or London. At times, the opposite seems to have occurred—the work discipline of the white German colonial officer, his need for recreation and work breaks in a treacherous climate, came under scrutiny. Archival silence should not lead to the conclusion that time measurement and distribution did not also serve the surveillance and control of colonized societies, although given the unstable nature of time, the degree to which official time could attain such a function is questionable.

The main proselytizers of European time appear to have been missionaries and employers of indigenous laborers. In recruiting Africans and others for excruciating labor on the sugar cane plantations in Natal or the Caribbean, rubber in Southeast Asia, and mines in Southern Africa, Europeans subjected colonized peoples to harsh regimes of discipline and punishment. With the onset of the so-called scramble for Africa, the abundance of cheap labor had been touted as the great boon of colonialism back home in many European countries. When the so-called natives refused to abandon their own subsistence economies in favor of European wage labor, their colonial rulers utilized a number of measures designed to drive Africans into seeking employment. Dispossession and expropriation, excessive hut taxes, and marriage fees—the latter to be paid only in the currency issued and used by the colonizer—were among the most notorious methods of coercing indigenous populations into wage labor. Unwillingness to carry out certain tasks that stood in conflict with the cultural, social, and religious beliefs of colonized peoples was read as general indolence. Non-Western societies were stigmatized as lazy, and corporal punishment was par for the course.[49]

Conflicts over time moreover emerged from variations in notation and computation. The Zulus in the British Natal Colony followed the moon and the stars in calculating the month and divided the year into thirteen moons. One circuit lasted for about twenty-eight days after which an interlunary period followed, a moonless period during which people paid respect to the darkness and abstained from work. Colonial officials, employers, and native workers clashed frequently over when a worker was due his pay, as European months followed different calculations.[50] Besides employers, it was above all colonial missionaries who longed to convert indigenous populations to Christian routines and rituals. Missionary schools like the one run by British missionaries in Lovedale in the Cape Colony were governed by a tightly hewed regimen of work and study without abhorred "vacant" time. By exporting their methods of organizing the flow and the use of time, missionaries intervened deeply in indigenous languages and imaginations. Jean and John Comaroff have shown how British missionaries systematically sought to transform the basic categories of indigenous existence among the Tswana in southern Africa, such as work, housing, and dress. Reorganizing indigenous time was always part of such endeavors. The goal was not least to give native daily life a rhythm and structure to begin with, thus breaking the monotonous cycle of eternal return characteristic of "primitive" societies.[51] Given the partial and unstable application of the official time of the colonial state, the brutality of colonial labor regimes and colonial schooling appear to have constituted at least initially a more grave assault on the real and imagined times of non-Western societies.

• • • •

BESIDES THE SLOW EMBRACE of uniform times by the colonial and non-Western state, conflict was another source of temporal instability and variety. In the first half of the twentieth century, war, occupation, and subsequent independence from alien rule were powerful agents of time change, daylight saving and other. Bureaucratic political time was adopted late only to be changed repeatedly. During World War I, Greece had introduced Athens time as the country's mean time in 1916, a time roughly one hour and thirty minutes ahead of the time at the Greenwich meridian. After the war in 1919, Greece switched to UTC + 2. A few months later, the British High Commissioner in Constantinople announced that Turkey (nominally still the Ottoman Empire) was bound to introduce summer time for the current year and voiced his desire to see Greece do the same. The high degree of interaction between allied forces in a region still at war made a common time advised for coordination purposes. Greek authorities refused, enlisting the help of the Greek National Observatory in crafting a response.[52] At a moment when Greece and Turkey fought over influence, territory, populations, and boundaries in the Eastern Mediterranean, political reasons surely did their part in convincing Greece to reject "Turkish" summer time. Barely two weeks later, Greek troops invaded and occupied Smyrna on the Aegean coast of Anatolia. Home to a large Greek Orthodox population, the city served to corroborate Greek claims to Ottoman-Turkish territory, aspirations that had been nurtured by allied wartime promises for territorial gains. Roughly a year later, with the Greek occupation of Smyrna still in place and territorial claims unsettled, the British High Commissioner in Constantinople again ordered for summer time to be observed. Greek authorities in Smyrna refused to carry out his instructions, citing inconveniences to the inhabitants of the city.[53]

Time zones and mean times occasionally assumed ostentatious political meaning reminiscent of French rejections of "British" time. As victims of Japanese imperialism, Koreans came to experience this potential for evoking identities and power at first hand. With the Japanese annexation of Korea in 1910 came Japanese time (later, in the context of a different occupation, Japan would replace Chinese time in Manchuria with Japanese time). In 1954, under the scientific leadership of Korea's foremost astronomer, Lee

Won Chol, South Korea decided to introduce "Korean Standard Time," at UTC+8:30 lying squarely between China (UTC+8) and Japan (UTC+9). To Koreans, Japanese time was a "relic" of the occupation.[54] Previously, the application of Japanese time in Korea had been upheld in the interest of the American military stationed in Korea. As a British diplomat observed, America had "finely disregard[ed] both the rights of a sovereign nation and the facts of geography" when it rejected "Korean Standard Time" on the grounds that it would complicate interactions with the American headquarters in Tokyo. "It is surprising," the British commentator continued, "that Americans, used as they are to various time zones in their own country, should here have shown something of the inflexibility of the old-style Japanese militarists." The US Army finally yielded, letting Korea demonstratively shake off the time of Japanese colonialism. In 1961, the switch was reverted and Korea returned to UTC+9, albeit under the name of "Korean Standard Time."[55] China itself, after gradually applying UTC+8 in coastal cities, divided the country into five time zones after the revolution of 1911. With the founding of the People's Republic in 1949, however, China abolished these zones and created one time zone for the entire country, which is still observed today. Given the longitudinal extension of China, this is arguably the most extreme example of nationalizing time in a globalizing world.[56]

The upheavals of the first half of the twentieth century triggered such frequent time changes that authorities failed to follow closely. In 1919, the Colonial Office made a perfunctory attempt to require colonies and protectorates to report unilaterally adopted time zones to London, but little seems to have come of this effort. Four years later, the International Telegraph Bureau in Berne contacted Britain's General Post Office in an attempt to compile a list of times in use around the world. For certain British colonies and dominions, Berne held, that information had proved hard to come by, but surely the Post Office would be able to help out. The General Post Office had to turn to the Colonial Office to collect the data to be passed on to Berne. The Colonial Office was unable to retrieve the requested information and simply began collecting bits of information here and there. Mostly, the solution was to write to the administrations in Britain's overseas possessions to obtain the information necessary for completing the picture.[57] Even the time department at the Royal Greenwich Observatory professed to being overwhelmed. "Because of the frequent alterations in standard times and the lack of certainty of information about such alternations [*sic*] being received," the staff declared itself incapable of providing up-to-date information. Neither was there a centralized system of reporting and documenting time changes nor a particular drive toward uniformity.

Britain's handling of time across the empire resembled what has been characterized as the "unfinished" nature and "project" character of said empire. "British expansion was driven not by official designs but by the chaotic pluralism of British interests at home and of their agents and allies abroad," is how one historian characterizes imperial politics. Though smaller in size and less diverse, the fabric of other European empires and colonial territories perhaps did not look quite so different.[58]

••••

ONLY IN THE 1950s did the dust settle on frequent time changes. In the 1940s and 1950s, a few spots continued to hold out on local times—Aden and British Somaliland used UTC + 2:59, Calcutta was still on its local time five hours and fifty-three minutes ahead of Greenwich, and many countries had yet to abolish mean times with half- and quarter-hour differences. Oddities remained. In Guayaquil, the capital of Ecuador, clocks ran five minutes and seventeen seconds behind the time in the rest of the country.[59] Gradually, however, countries new and old settled on mean times and summer times that no longer differed by half-hours, quarter-hours, or just as common, minutes from Greenwich time. By the 1950s, finally it was possible to speak of a system of uniform, mostly hour-wide time zones girding the globe, with a handful of exceptions still in place. A mix of factors seems to have propelled this last wave of unification moves. War had been an engine of standardization in many technological regards, as war production required a tightly organized regime of industrial standard setting. Following World War II, the International Organization for Standardization (ISO), founded in 1947 in Geneva, promoted industrial and commercial standards at the height of Fordist production. Such efforts meshed with national policies especially in the United States, where standardization was a key concept to economists, engineers, and others. Magazines reporting on the latest developments in industrial standard setting cropped up. One possible corollary might have been a greater insistence on time standardization among the Euro-American development officials, military advisers, and modernizers who now populated the emerging third world on a mission to dole out development aid and impart expert knowledge.[60]

For several decades, the practical need for a uniform system of time zones was not pressing enough to convince governments at home and abroad to push for its implementation. The perception and ideology of a globalized world in which everything and everybody was connected to everybody else stood in contrast with the deliberately regional and national solutions still so dominant in timekeeping. It was entirely possible to remain outside of unified time without, it seems, incurring greater political and other losses. This changed only gradually. Whether or not it is indicative of a more general trend, Imperial Airways, predecessor of today's British Airways, wrote to the Astronomer Royal at Greenwich in 1937 and requested information on the times kept in a handful of more remote locations, among them Basra in Iraq, Sharjah, Luxor, Mombasa, and Beira (Portuguese East Africa).[61] The rising interconnectedness of the world had hitherto been captured by using the word "world" to form new expressions or, in some languages, words and concepts that conveyed awareness of a globalizing world. "World time zones," "world calendar," and talk about a "world language" (Esperanto) are only a few such examples. Beginning in the 1940s, "global" as an adjective slowly began to take the place of "world." Publications now containing the word "global" in the title often dealt either with the war or aeronautics, or both. Titles such as "From Global War to Global Peace," "Global Mission," "Global Geography," "Global War: An Atlas of World Strategy," "Our Global World: A Brief Geography for the Air Age," "India and the Global War," seem to suggest a confluence of defense and military interests with civil geography and aviation. It lies beyond the scope of this story, but such an amalgamation of strategic and aviation interests in the jet age may have ultimately created a practical demand for more unified times from which it was now harder to abstain.[62]

The spread of zone-based mean times around the world brings to light the extreme unevenness of globalization. The application of territory-wide mean times was not only protracted and spotty; it also percolated without direction, often at the initiative of local officials with local and national (at times regional) goals in mind. In this, the convolutions of uniform time resembled time's circuitous journey within states such as Germany or France, where single regions and cities took steps toward introducing new mean times that preceded the eventual legalization of nationwide time changes. Similarly, scientific associations and societies operated as vectors of time transmission by alerting colonial and national government officials to the possibility of introducing uniform time, just as they had previously done in Europe in the 1880s and 1890s. The arguments wielded by the members of the countless amateur scientific, geographical, and historical associations that sprang up in the nineteenth-century world with the goal of convincing those in power of the need to unify time manifested the ideology of globalization: the invocation of a shrinking world of instantaneous communication, rapid transportation, and universal interconnectedness. The long laborious and concrete road toward a grid of twenty-four mostly even-shaped zones of abstract time, and the frequent, erratic switches originating with local administrators and legislatures, stand in stark contrast with the perception and invocation of interconnectedness. Uniform time zones mostly based on Greenwich spread eventually, after national and regional integration concerns no longer stood in contrast with global considerations. Only after a combination of modernization drives, military, and aviation interests made it increasingly cumbersome to remain outside a now widely applied system did time unification become a reality. But by mid-century, the internationalist movement that had pleaded for uniform time in the interconnected world of the late nineteenth century had long since faded. Uniform time succeeded without internationalism at its side. The allusion to the supposed prerequisites of an interconnected, shrinking world that mandated uniform time was by and large a way of interpreting that world from the perspective of Europe and North America, a worldview germane to the second half of the nineteenth century.

CHAPTER FOUR

# A Battle of Colonial Times

THE UNEVENNESS of globalization was manifest not alone in the spread of uniform time zones and mean times. Uneven globalization also resulted in the coexistence of "reformed" and archaic times, and as observed in the case of summer time, of time discipline paired with an adherence to natural and biophysical rhythms. In the non-Western world, such composited times often resulted from an amalgamation of uniform, reformed times and local beliefs and interpretations of time. Responses, reinterpretations, rejections, and adoptions of Euro-American time in non-Western societies were shaped by the fabric of society and political organization that Europeans encountered or sought to engineer upon landing on far-flung shores. Since the duration and intensity of European colonial rule varied considerably from place to place, the means and institutions through which local societies were subjected and exposed to alien concepts of time differed accordingly.

Those colonial societies where educated colonial subjects had been exposed to European curricula and now worked in lower-rank positions in European administrations, offices, and businesses opened spaces for engaging with varieties of time, not always on appreciative terms. The colonial workplace was one site where the stakes of adopting official mean times became apparent. The politics of nationalism was another. In one such colonial context in India, the British presence was more than a century old by the late nineteenth century and had thus generated an English-educated middle class that cautiously began to question various aspects of the Raj. In British India and Bombay in particular, the response to British plans for adopting Greenwich time for the subcontinent was therefore fueled by a particular fusion of local and national, Bombay and Indian identities.

British officials may have painted refusals to submit to British time as manifestations of insurmountable backwardness. But Indian insistence on applying local Bombay time instead of the time of the colonizer became a token of national "Indian" collective identity and thus an important feature of Indian political modernity. The rejection of the colonizer's time constituted an attempt to create histories and identities that were not calculated with reference to the British temporal measure of all things and to reject India's assigned position on the global map of universal time.

• • • •

By the end of the nineteenth century, Britain's India had become a linchpin of empire. Economically, British surplus balances with India offset the growing deficits with Germany and the United States, making Britain and India the pivot of a complicated system of multilateral relations. Militarily, the Indian Army advanced to a substantial strategic reserve not only for securing British rule on the subcontinent but to protect her interests in all of Asia if need arose. Financially, Indian earnings helped free up resources in the City of London to be invested elsewhere and moreover facilitated the London-centered monetary system of the gold standard. And yet, even here, at the heart of the British Empire, the colonial state only late and reluctantly unified the several clock times in use under its rule.[1]

If colonial British officials were slow to act, scientific organizations and associations, increasingly well connected outside of Europe and North America, eagerly filled the void. Only when prodded by these organizations did the British Government of India move to operate a single time in what was arguably a cornerstone of empire. Once it did so, the reactions among some of its subjects were hostile. The introduction of a mean time for all of India coincided with other largely unpopular measures enacted by the British in India. A new time fueled outrage among early Indian nationalists who were cautiously beginning to picture future Indian self-government. One city in particular became the stage for a prolonged conflict between British authorities and colonial society: Bombay, a commercial and intellectual hub perched on the western coast of the Indian subcontinent. Time reform, and the encounter with a time imposed by the British imperial rulers, rendered visible the interplay of scales between local urban identities, global integration, and the forging of national and nationalist communities.

The city of Bombay had undergone deep transformations in the second half of the nineteenth century. Events that played out far beyond the subcontinent had an important share in these changes: during the US Civil War, cotton production and export came to a grinding halt in America, thus forcing the British textile industry to ensure its supply of raw materials elsewhere. British India became one of these outlets, and Bombay, gateway to

India on the western coast, handled the bulk of cotton exports. The Civil War ushered in a veritable boom for Bombay. While the upswing did not last, the city's merchants quickly found other profitable enterprises, benefiting from previously completed infrastructure projects on the subcontinent as well as globally from the opening of the Suez Canal. At a stroke, the new thoroughfare reduced the travel distance between Bombay and London by over 40 percent.[2]

Bombay's remarkable commercial rise was accompanied by a change in the social and intellectual fabric of the city. Its geographical location had always made Bombay a hub for a variety of trading communities, qualities that were only enhanced by Bombay's growing importance to the British system. The inflow of capital drew wealthy entrepreneurs from adjacent areas as well as migrant laborers from more distant parts of the subcontinent to the city. The city's already variegated religious landscape, consisting primarily of Hindus, Muslims, and Parsis, similarly became even more diverse in the course of such movements. These ties, sustained not least by improved communication and transportation, meant that "by the nineteenth century, Bombay's economic and political influence was unmistakably national in scope."[3] Exuding confidence, the citizens of Bombay thought of themselves as harbingers of cultural and political trends as well, claiming for their city the status of *Urbs Prima in Indis,* India's first city. The city's vibrant scene of local newspapers in vernacular languages and several associations and civil society organizations often hosted spirited debates about politics and intellectual affairs and together generated a lively public sphere.[4]

It was in this city that, for the second time, a missed train made history. In 1881, the British governor of Bombay, James Fergusson, failed to calculate his way through the thicket of simultaneously existing different times and schedules applied on railway lines, the telegraph, and local town hall buildings throughout Bombay. Fergusson made it late to the station, and when the train departed without the governor on board, he was thought to have developed a "spleen" about all things time.[5] Out of this obsession an idea was born to unify the times in use across the region. Within the greater administrative unit around the city of Bombay, the Bombay Presidency, several times were kept: Bombay and Poona each had their own local time; Ahmadabad followed Madras time. Fergusson's move would serve a twofold goal: to overcome the difference between railway time and the times otherwise followed in Bombay, and the discrepancies within the presidency at large.[6]

In Calcutta, Fergusson or any traveler would have encountered a similar potpourri of times as one advocate of more uniformity characterized the persisting problem several years later. "We should no longer have the clock outside the General Post Office pointing to one time, and that on the Howrath platform pointing to another," the complaint stated, referring to Calcutta's main train station. Under a more uniform regime of time, the report went on, "the traveler would no longer have to make an intricate calculation to find out at what time (local) he would have to leave his house to catch a train which departs at another time (Madras). If he set sail for Burma, or went up the river to Assam, he would not need, on arrival at his destination, to make anxious enquiries as to the time in use there, for it would be exactly the same as what his watch showed; and if he traveled in the other direction to Madras, Bombay, or Delhi, he would only have to remember that the time there was exactly an hour slow of his watch."[7] In a similar mélange of coexisting clock times in Bombay, Governor Fergusson simply forgot to account for the difference between his city's time and railway or Madras time.

In these decades, the Scottish-Canadian engineer Sandford Fleming, one of the initial promoters of the global time zone system, was frequently speaking at conferences and congresses, praising the concept of time zones by relating an episode in which he missed a train in Scotland due to the unclear relationship of timetables to local time. Fleming's experience might or might not have informed any of the urban legends spun about the figure of the Bombay governor. In any case, on November 8, 1881, Fergusson put his plan to work and informed the citizens of Bombay "that from and after December 1, 1881, Madras time shall be kept in all offices under the control of Government and shall be held to be the official time for all purposes."[8] Beyond governments and railways, for the sake of uniformity, the public was invited to accept Madras time as well. Madras time was roughly forty minutes ahead of local Bombay time. When smaller local and regional railway lines began sprawling and connecting across the subcontinent from the 1850s, it became necessary to forge an overarching time standard for the main branches. Without much notice taken by the general public, Madras time had been used on telegraphs since 1862 and railways beginning in 1870.[9]

The 1881 introduction of Madras time to the city of Bombay came as a complete surprise to the local populace. Newspapers were quick to reject the measure. "Without good reason," in one paper's words, summed up the reactions. Another outlet opined that while Madras time perhaps could be reckoned in offices, the city's public clocks surely would have to remain on local time as nobody could imagine the replacement of true solar time by a new time standard. Bombay denizens, both British and Indian, began writing letters to the editors of some of Bombay's most widely read periodicals in both English and Indian languages. One recurring point of complaint was the manner in which the new time altered the relationship between rhythms of everyday life and daytime and darkness. By moving the clock forward under Madras time, office clerks were forced to rise half an hour "earlier," left with the choice to either get up in the dark or hasten through a morning meal and the daily commute. Madras time had the effect of reducing the time between sunrise and the beginning of work by half an hour. Such complaints prefigured debates about daylight saving that would torment British observers in Europe roughly twenty years later.[10]

Other criticism was directed at the disorder and misunderstandings caused by the time change. Writing under the pseudonym "Anti-Despotic," one reader lamented that prior to the introduction of Madras time, comparing times had only been necessary for rail travel. Now, comparison was required in every moment of life. Even a simple appointment between two people forced the question, "Supposed I submit and keep Madras time, how am I to know that Jones and Smith do the same?" Another reader described a gathering of members of the Bombay municipality, a local government body, to which attendants arrived half an hour late or early due to the switch of times.[11] "Confusion" and "misunderstanding" were the most frequently used terms to describe the situation that emerged in Bombay after November 1881. The government dockyard and the police office kept the new Madras time; the clock of St. Thomas Cathedral had been switched to Madras time; in private offices, schools, even at the High Court, Bombay time remained in use. "Consequently great inconvenience is felt by a large number of people," the *Bombay Samáchár* wrote.[12] As in Europe, some chose to imagine the time change as merely an alteration of nomenclature. "What we call now six o'clock we are asked to call in future 'half past six,'" a reader of one of the British papers in India explained. Such an amendment would require no change of habits at all. Those arriving at work at ten would still go at that moment (in "true" solar time) and only speak of it instead as 10:30.[13]

In Bombay as in Paris and Berlin, old habits died hard. Governor Fergusson as much as other bureaucrats of the nineteenth century learned the hard way that decreeing the application of a new time did not guarantee it would actually be reckoned with. In the years after Fergusson's contested measure was passed, the discussions and complaints never fully came to rest. It was not without schadenfreude that the *Bombay Gazette* remarked as late as 1883, "it may be mentioned that in none of the Bombay Clubs has Madras time been introduced."[14] Public time was no better off, as another article noted. "The many clocks which are now open to the public view are rendered useless for public information, because they are all at sixes and sevens, and the usefulness of the true time clocks is nullified by the false time clocks, except to a few persons who happen to know which time is kept by any particular one."[15]

The clock at Bombay's university became a particularly contested object in the struggle between the authorities and local citizens. The tower clock had traveled by ship from Britain and was installed in 1882, but the university soon realized it lacked the funds to cover the cost of lighting the clock faces at night. A fund-raising effort brought the Town Council on board but its members explicitly stipulated that the clock show Bombay time. Governor Fergusson begged to disagree and refused to authorize funds for a clock displaying anything other than "his" Madras time. He was, however, willing to foot the entire bill for lighting a clock set to Madras time. In April 1883, the Bombay Chamber of Commerce, itself in favor of Bombay time, successfully goaded the university senate into holding a referendum on the question of displaying Madras or Bombay time.[16] With overwhelming majority, the body opted for local Bombay time, causing the municipality to immediately cut financial support for maintenance and illumination for a clock giving "unofficial time." Government authorities did not tire from repeating, "since the clock will keep Bombay time, not a rupee shall be contributed out of the provincial funds to enable the citizens to see that it does not keep 'official' time."[17]

April 1883 was the high water mark of public dismay over Bombay times. Newspapers incessantly demanded that local Bombay time be reinstated until Governor James Fergusson finally succumbed to the prevailing public pressure. Attempting to save face, Fergusson tried to lay the blame at the feet of the Chamber of Commerce. It was now said to have been the Chamber that pushed for the introduction of Madras time in 1881; now, two years later, that very same body had undergone a change of heart. Since the Chamber no longer believed in the benefits of following Madras time in Bombay, the government in May 1883 was simply following its lead in ordering a return to Bombay time. Wherever the blame fell, in the press, Fergusson's return to reason was celebrated as a victory of public opinion over the government. Between 1881 and 1883, the British government's attempt to switch from local to railway time elicited opposition from both English and Indian inhabitants of the city. What shaped these responses was a strong local urban identity that refused to accept the time of a rival Indian metropolis as the new mean time.[18]

• • • •

FOR THE FOLLOWING fifteen years, official times remained largely unaltered throughout the subcontinent. Times remained variegated, with Madras or railway time applied sporadically. A report from 1903 estimated that

Madras time was followed to some extent in the Punjab, Baluchistan, Sindh (except for Karachi), United Provinces, and the Madras Presidency. Burma observed Rangoon time, and the rest of India recognized some version of regional or local time.[19] Only when scientific associations and organizations in Europe began to rely on their increasingly far-flung network of branches and connections for information about applied times, did the question of time unification return onto the agenda of colonial administrators in British India and elsewhere. In October 1897, a professor and member of the Seismological Investigation Committee of the British Association for the Advancement of Science approached the Colonial Office in London with a request. The initiative passed through various hands until it ended up before the Government of India. As the scientist explained, he was preparing a publication that would list the "differences of local time and Greenwich Mean Time in various parts of the world," an instrument he hoped would prove indispensable to the precise prediction of earthquake times. Such and other lists helped publicize the "state" of timekeeping throughout the British Empire and raised awareness for the current disarray. One year later, in 1898, writing in a similar vein, the same correspondence by the Royal Scottish Geographical Society (RSGS) that had ushered in the adoption of mean times in East and Southeast Asia reached the India Office in London. India would become another locality where time was changed at the initiative of a scientific association.[20]

The RSGS declared that "in these days of rapid travel and incessant interchange of telegraphic news between distant parts of the world, the desirability of establishing a general 'Standard Time' has become more and more apparent."[21] In arguing for more uniformity, the Society pointed to a deplorable irregularity that was unworthy of an imperial power of Britain's stature: the current edition of the "Admiralty List of Time Signals" of 1895 listed 153 stations distributing time signals for ships and navigation. While 94 stations emitted signals in Greenwich time, the remaining 59 based their emissions on a different meridian and mostly on a local time. To the Society's unmitigated dismay, 21 of these aberrations emanated from British-owned soil beyond Europe and North America. To end this regrettable state of affairs, the introduction of Greenwich time in British possessions overseas was strongly recommended. Extending British time to its overseas possessions would moreover spell "the abolition of the present barbarous arrangement, unworthy of a country pretending to civilization, by which every place keeps its own time."[22] What is more, as others pointed out, even after Madras time was officially adopted for railways and telegraphs in India, train schedules and the official telegraph guide continued to print time differences between local and mean times, stretching over as much as forty-one pages, and, as one observer remarked, directly encouraging the "maintenance of the present inconvenient and antiquated system," constituting "a hindrance to the adoption of a more rational one."[23]

Less unanimity existed on the question of which time difference to Greenwich the new time was to observe. The Royal Scottish Geographical Society suggested adopting the time five hours ahead of Greenwich for the entire continent.[24] The Society's main London branch, however, while siding with the RSGS's general calls for Indian time reform, proposed a different format. It acknowledged the desirability of establishing hour-wide time zones and differences yet cautioned, "it would be a serious mistake to attempt to enforce any arbitrary uniformity of system that failed to comply with the essential condition of practical convenience." Instead, as it explained, the meridian five hours and thirty minutes ahead of Greenwich neatly divided the subcontinent into two almost equal halves.[25] Briefly, members of the well-connected scientific societies even looked at America and considered a system of multiple hour-wide zones for South Asia.[26]

For the time being, the combined efforts of learned societies and associations failed to convince colonial officials to initiate time unification. George F. Hamilton, secretary of state for India, replied, "The time has not yet arrived for action such as suggested by the Societies." Moreover, he argued, with Madras time, a uniform time was already in use. The secretary foresaw "considerable practical difficulties in prescribing it [a new mean time] at places like Bombay, Calcutta, and Karachi." India also extended over such considerable longitudinal distances that a single meridian would prove difficult. "It may be admitted that a change will some day be inevitable. But in the meanwhile it seems desirable to watch the working of the different systems, and when the time for a change arrives the experience of other countries will be available as a guide to the ultimate decision," his reply closed.[27]

• • • •

IF OFFICIALS in London and Calcutta hesitated to impose a change of time, reform-minded scientists did not relent. Again in 1902, another association, this time the Observatories Committee of the Royal Geographical Society, addressed the colonial government on the issue of a mean time for India. Upon receiving the renewed correspondence, John Eliot, a meteorologist in the service of the Government of India and director general of Indian observatories, now sat down to assess the totality of different proposals received over the past years. Eliot concluded that a switch from Madras time to a time standard of five hours and thirty minutes east of Greenwich would constitute an improvement, but that with regard to international scientific purposes, two times of five and six hours ahead of Greenwich each would be preferable. Yet despite this assessment, Eliot changed his mind and ultimately endorsed a one-zone solution: a mean time of UTC + 5:30 would "begin to bring India into line with the rest of the world, as by far the great majority of civilized countries have adopted international time in one form or other." A well-oiled rumor machine later attributed the about-face to pressure from various British-operated railways weary at the prospect of operating more than one time, especially where railway lines potentially crossed from one time zone into another.[28]

In yet another show of hesitation and reluctance, almost a year passed until the government moved. In July 1904, provincial administrations and railway companies received a circular by the Government of India with an enclosed anonymous position paper titled "Note on a Proposal for an Indian Standard Time."[29] The government asked recipients to furnish their views on the proposed change, thus allowing authorities to gain a more complete picture of possible obstacles involved. The memorandum itself spoke the language of the global imagination when stating, "in these days of universal international transactions, of rapid and extended communications, and especially of almost instantaneous telegraphy, every country is concerned with the time of every other." Already, the note stated, "ships, all the world over, set their chronometers to Greenwich time, and it is an intolerable nuisance to have to take a pencil and a piece of paper in order to compare their time with the local time of the port in which they happen to be." The position paper came out strongly in favor of one single time zone, as two zones would constitute "a retrograde step" in comparison with the already existing system of uniform Madras time.[30] Although it was the existence of a globalizing world that mandated time reform, the solution favored was clearly moved by "national," or more appropriately, India-wide concerns. Contemporaries saw no contradiction between this solution and universal integration. It was a logical conclusion to the status quo.

Responses from provincial governments, railway authorities, and associations of different shade came trickling in during the fall of 1904. Local governments had mostly mailed inquiries to merchant communities within their purview and now enclosed the feedback with their responses. The Government of India had deliberately left open the question of whether the new uniform time would find application only in matters of transportation and communication, or whether it would be extended to cover all aspects of civil life. Among commercial associations, the mood was amenable to the proposed change. The Karachi Chamber of Commerce provided a list of twenty firms that had spontaneously articulated their consent.[31] In the Madras Presidency, the Harbor Trust Boards, the Port Officer, the Chambers of Commerce in Madras and Cochin, and the Madras Trades Association had all given the project their seal of approval.[32] Yet the Bombay government initially urged caution, for it found, "there is a feeling in the City of Bombay and an evident though less marked feeling in Karachi in favor of the retention of local time for general purposes."[33]

In the face of what appeared to be a solid consensus among those canvassed, the Government of India moved to introduce the time five hours and 30 minutes ahead of Greenwich to the colony. The new time was designated as "Indian Standard Time," to deflect from its potentially controversial "British" source. In a curious borrowing, the "standard" in Indian Standard Time appears to have been derived from the initial correspondence among popular-scientific associations about time zones in the American context. American "standard" railway time thus became Indian Standard Time. Hour-wide zones were brushed aside by overriding imperial and "national" Indian concerns. Since a half-hour difference was deemed most suitable by officials on the spot after all was said and done, it was introduced based entirely on political and economic concerns derived from British interests in India, not with regard to a worldwide system.

In January 1905, the Government of India instructed the Public Works Department to introduce the time five hours and thirty minutes in advance of Greenwich as "Indian Standard Time" while in Burma the time to be adopted would run six hours and thirty minutes fast. As of July 1, 1905, all railways and telegraphs on the Indian subcontinent were to follow the new time.[34] In order to attenuate dissatisfaction about altered relations of daily life to daylight, the Bombay government amended a range of official hours throughout the city; other institutions followed. The Port Trust, the city's time balls, the Associated Exchange Banks, all changed opening and work hours to retain the same relation to solar time as previously. As the new mean time differed from solar time by roughly thirty-eight minutes, the adjustment sometimes involved a half-hour, sometimes an hour change.[35] Often, such a move was justified by British officials as meeting the needs of the "native" population, which according to this argumentation, was unable to live under any other time than true solar time. Against the backdrop of near simultaneous discussions about summer time in Britain and the striking inability to think of time as severed from natural rhythms, it is more likely that such recalibrations ultimately reflected the British rather than the "native" imagination of time.

Critically, however, authorities made no plans for making the extension of the new mean time mandatory beyond railways and telegraphy. It left the decision to local and regional authorities and added, should "the general public in centers such as Calcutta, Bombay, and Karachi, which at the present follow the local time of their respective longitudes, evince any desire to adopt the new standard for daily use in place of local time, we are prepared to give our best support to the change by ordering the adoption of the new standard at Government institutions in those cities." Aside from such assistance, the colonial state refrained from actively converting local times into colony-wide mean times.[36]

The Government of India accurately anticipated the opposition to uniform time it was about to unleash, although perhaps less so its scope and intensity. Once more, it was Bombay, and to a lesser extent Calcutta, that became the focal point of collisions between deeply rooted urban identities and imperial policies. In 1905 as compared with 1881, protests against a new colony-wide mean time struck a much more anti-British chord than previously. Now it mattered that this was a time decreed by the British colonizers, that it was "British" time being imposed on colonial subjects. Twenty years after the Government of India's first brush with time, under the changed circumstances of British rule in India in 1905, retaining local time became a matter of Indian national politics.

Indians now perceived the change in official mean times as yet another in a long series of attempts by the colonial state to meddle with local and personal affairs. Following the repression of the Indian rising in 1857/1858, the colonial regime that replaced the East India Company's rule began to slowly expand the scope of the colonial state. At the same time, in a move characterized more by paternalistic notions than a desire for more participatory colonial politics, the British in India since 1892 widened the responsibilities of local political bodies while opening membership for select positions to Indians. A growing number of Indians who had been educated in British schools and universities on the subcontinent moreover formed their own local and regional political and civil society associations. In 1885, in this case prodded by a British civil servant, the Indian National Congress was founded, taking its seat in Bombay. In such and other associations, British-educated Indians were given a moderate arena in which to engage in an initially British-guided dialogue with the Raj. By 1905, the atmosphere of both local and national politics was therefore more politically charged than twenty years earlier.[37]

Since the short-lived experiment of running Madras time throughout Bombay in 1881–1883, time had largely remained as multifarious as before. The city continued to use its own local time while railways followed Madras time. A time ball erected at Bombay Castle was dropped at 1:00 p.m. local time, providing orientation to the broader population. Another time ball fell daily at 12:00 p.m. at the Prince's Dock, primarily serving ships and sailors in the port. At Bombay Castle, the seat of the government, another time signal was given in local "true time" at 1:00 p.m.[38]

Such was the situation when Indian Standard Time was to be introduced in the summer of 1905 on railways and telegraphs. Emboldened perhaps by similar moves of other local administrations, Bombay authorities suddenly made the decision to push for the adoption of Indian Standard Time for all official purposes and in government offices throughout the Bombay Presidency. In October 1905, the Government of Bombay asked the Government of India for permission to introduce the new time. Several local associations, the Bombay government wrote reassuringly, had declared their support for such a step. The port administration, the Bombay Presidency Trades Association, the Bombay Association of Fire Insurance Agents, the Native Share Broker's Association, and the Mill-Owners' Association had replied positively.[39] The Bombay Chamber of Commerce had begun to explore the question of time unification already around 1903. Once the Bombay government had reached its decision, the Bombay Chamber summoned a special session to cast a vote on the government's announcement to apply Indian Standard Time as of July 1905. In a surprise result, however, the Chamber wound up opting to retain Bombay time in a 22 to 17 decision. That settlement was upturned shortly thereafter. At the behest of some of its members, the Bombay Chamber of Commerce called another meeting on standard time. The reason was that at the first session a disproportionately small number of members had been present to cast a vote. Now at the second show of hands, the Chamber sided in favor of the new time in a vote of 51 to 16.[40]

Since many of these associations were made up of a significant number of Indians, British officials apparently concluded the Indian population was accepting of a new mean time for Bombay. But they forgot to reckon with one institution that would become a bastion of public resentment. In October 1905 and again in December, the Bombay Municipal Corporation, the local self-governing body of the city, had declared itself "in favor of the adoption of Standard Time in the city" and was "prepared to adopt it for all Municipal purposes." In January 1906, the Municipal Corporation did just that and resolved to follow standard time in Bombay—carrying the vote by one voice.[41] It was later reported that the European members of the Municipal Corporation rushed a vote by crying "vote, vote" after one Indian proponent of the time change had spoken in favor of mean time, thus preventing others eagerly positioned in the starting blocks from offering a "crushing rejoinder" to his speech. European members normally did not even attend meetings regularly, critics were fast to point out; but on the day of the standard time vote they showed up as a "phalanx." Fourteen of the yea-sayers were moreover from among the European representatives who were appointed to the Municipal Corporation by the Government of Bombay and hence "bound to vote for the government measure, never mind what their conscience dictated." The article hurling these charges identified further voters who in one form or another could be counted as "official" and therefore not impartial representatives. All that was left was therefore a small group of nine truly independent voters who had spoken in favor of adopting India-wide time in Bombay.[42]

Far from over, debates now rapidly grew more heated. Three months later in April, the Municipal Corporation was forced to take up the matter anew, pressed by Bombay citizens who had drafted a petition that was successfully brought before the corporation. Several attendants gathered at the session of the Municipal Corporation mostly to view Pherozeshah Mehta take the stage. Mehta, a member of the Bombay Parsi community and also known as the "Lion of Bombay," was a prominent figure in local Bombay politics and early Indian nationalism. Mehta was involved with drafting the 1872 municipal reform act aimed at granting Indians a modicum of participation in local politics, and later was a member of the Bombay Legislative Council as well as the Imperial Legislative Council. As one of its founding members, Mehta also presided over the Indian National Congress in 1890. Pherozeshah Mehta turned into one of the most vocal opponents of applying Indian Standard Time in Bombay.[43]

At a renewed meeting of the Bombay Municipal Corporation in April 1906, Mehta introduced a proposition to return to local Bombay time. His arguments were trenchant. "It is not fair and proper that the population of this City should be driven like a flock of dumb cattle because the Chamber of Commerce and the Port Trust adopted Standard time. . . . a measure adopted by Government without consulting the feelings and sentiments of the people and without giving them an opportunity of expressing their opinion," Mehta contended.[44] "Standard time never could be adopted in Bombay except by the small colony of Europeans and natives that go with them." The introduction of standard time in the city, he claimed, had led to several impracticalities and inconveniences. As of May 1906, all municipal clocks should therefore be reset to local Bombay time. After more than three hours of "heated discussions," thirty-one members of the Municipal Corporation voted for Mehta's proposal, twenty-three voted against it. As newspapers concluded, it was his "stirring" speech that convinced the members to return to Bombay time.[45]

Outside the meeting halls of the Bombay Municipal Corporation, the general public was voicing its dislike for the new order of time evermore loudly. As with other time changes in Europe and North America, the new

Indian mean time was criticized for being "artificial" and unnatural. "We are asked to forget our natural time, the same that we have been familiar with from times immemorial, and adopt the new 'standard' which the ingenuity of the Astronomer Royal has devised," the newspaper *Kaiser-i-Hind* complained, adding that nature herself must be in rebellion against this time.[46] Later, the paper proclaimed, "nobody has asked for artificial time" to replace a time "which Nature has given to us and which mankind has faithfully followed these eight thousand years at least." A letter to the editors of the *Bombay Gazette* found the new time to be "fictitious." Another newspaper established, "the solar time is really the true time which regulates the affairs of each Indian household." Pointing to the disingenuous nature of the new time neatly gave way to accusations of interfering with the religious practices of Hindus, Muslims, and Parsis alike, all of whom relied on solar times in one way or another for performing religious duties. Indian merchant associations and newspapers repeatedly addressed this concern.[47] Bombay's role as India's main gate for Western trade routes, its function as the port through which Muslim Hajj pilgrims from as far away as Xinjiang almost necessarily had to pass when embarking on their route to Mecca, had long since made the city a site of diverse religions. When a "foreign" time was to be introduced to their town, different religious communities often protested with one voice.

Criticism was moreover directed at the political economy of time unification. Already back in 1881, a serious concern among those rejecting Madras time was the impact of the new mean time on the working hours of salaried employees. Daily routines especially of the working population were regulated by the sun and closely correlated with the periods of daylight and darkness. The paper *Indu Prakásh* therefore surmised that the innovation would "work more or less to the prejudice of the native employés [*sic*] of Government by compelling them to attend their offices half an hour earlier as hitherto had been the case." The *Bombay Gazette* added, officials obviously took for granted "that all the employés of [*sic*] Government will be able to attend office half an hour earlier without inconvenience." To many it was evident that the weakest link in the chain of employment would be hit hardest, "the already hard-worked clerks will suffer by the change."[48]

Similar worries flared up anew when the Indian government announced the introduction of the new mean time in 1904. The Bengal Chamber of Commerce launched an inquiry into the question of opening hours and debated whether these hours had to be adjusted in offices and courts "in order to provide that the actual working day shall remain in the same relation to sunrise and sunset."[49] Occasionally, Indian newspapers fell into the same habit as European ones in imagining orders of time: they assumed an unchanging, fixed rhythm determined by the sun to which the human body and mind were chained. "To have to attend to business or to attend office at 10:00, 10:30 or 11:00 a.m. is a thing one is accustomed to," one paper wrote. "But to have to do the same thing at 10:39, 11:09 or 11:39 will lead to a good deal of irregularity and . . . mistakes," the *Gujaráti* warned.[50] In Karachi, office hours had "from time immemorial" been set from 10:00 a.m. to 5:00 p.m., and the new system that forced offices to open half an hour "earlier" was disruptive to the lives of the majority of Karachi citizens. It was the "already overworked clerical subordinate" who would suffer most.[51] While the Indian "clerks" arrived daily between 10:00 and 11:00 a.m., their British "officers" seldom showed up before 2:00 or 3:00 p.m. The new time discriminated against "the clerical staff of the office, who are deprived of all recreation." Some even suspected nothing short of a colonial conspiracy, accusing the government of attempting to sneak in "an extra hour or half hour of work on subordinates in Government offices and mercantile firms."[52]

Fueled by ongoing debates and lingering dissatisfaction, Bombay citizens eventually decided to take matters into their own hands. In January 1906, the Indian inhabitants of the city, this time under the leadership of another known political activist, Ahmedbhoy Habibbhoy, addressed the Government of Bombay with a ringing petition, emphatically demanding that Indian Standard Time be revoked. Officials turned a deaf ear to these voices and, unmoved, responded, "It is believed that the people of Bombay will become accustomed to the nominal change of hours . . . without any serious disturbances of their former habits."[53] They were off by several orders of magnitude. Newspapers soon reported that workers in factories and cotton mills surrounding Bombay had taken to protesting; for according to their calculations, the new time imposed an additional forty-five minutes of daily work on them. A feeling of deceit prevailed. On January 5, 1906, roughly 5,000 mill hands gathered and started pelting a factory with stones, adamantly refusing to return to their looms under the new regime of time. In an act of symbolic protest, the striking workers eventually smashed the clock at one of the largest mills in the area.[54] The owner of the cotton mill had already altered the working hours. To reflect the time difference between mean time and local time, the workday now ran from 6:10 a.m. to 6:10 p.m. instead of 5:30 a.m. to 5:30 p.m. previously. But only when a full return to Bombay time was promised did the workers resume work. The mill owners were left with no choice but to set back the clocks to Bombay time.[55] Slightly later in Bombay, a demonstration was held at Madhav Baug, the center of the Indian section of the city. The gathering against the Bombay government's decision to enforce standard time drew some 3,000 people. At this event, Balachandra Krishna, another local politician, rose to a forceful defense of Bombay time against the "smuggling, by official intrigue of 'standard' in place of local time." The protest resulted in a petition demanding once more the return to Bombay time. Another petition had previously collected as many as 15,000 signatures.[56]

Criticism of manipulated working hours in combination with large demonstrations easily gave way to other, more explicitly politicized arguments. With so many Indian citizens opposed to the change in time, it was an obvious question to ask who, then, stood to benefit from the proposed change of time. Once non-Indian and, thus, mainly British merchants had been identified as the main culprits behind the adoption of an India-wide mean time, Indian newspapers began to openly question how "representative" these circles were of Bombay and India. After news emerged about the Bombay Chamber of Commerce's change of course—first opposing Indian Standard Time, then endorsing it—Indian papers raised the question of who accounted for the majority of Chamber members. One of the papers, *Kaiser-i-Hind,* went on to provide the answer itself, stating, "Surely they [the members] are not permanent residents of Bombay. They are aliens and foreigners. . . . They are only birds of passage—a microscopic minority." If a plebiscite on standard time were held, it was posited, "it would be found that ninety-five per cent of the population disapprove of the new fangled time, and the bulk of this microscopic minority of five per cent is European."[57] This was an interesting choice of words as "microscopic minority" was frequently used by British officials to lampoon those English-educated Indians who began talking about self-government in these years. In this line of argumentation, the British rulers insisted that Indian society was cellular, fragmented into myriad castes, ethnicities, and religions. Anybody claiming to speak for the totality of "India" could not possibly be representative.[58]

Another writer with *Kaiser-i-Hind* criticized, "The European community ought not at all to be considered in this matter. Its population is fractional, and is besides ever moving and ever shifting. . . . What is it to them if it is solar time or standard time or golf or tennis time or whiskey time?"[59] The Grain Merchants' Association adopted a "counter-resolution" that "strongly disapproved of the resolution passed by the Bombay Chamber of Commerce" while emphasizing that the members of its own association were mostly "natives" belonging to the "trading and mercantile classes" of the city—hence, commercial interest groups nonetheless but with diametrically opposed views about the desirability of the proposed time change.[60] In the same vein, the Bombay Native Piece Goods Merchants' Association found "that the Chamber does not fairly and adequately represent native merchants" and came to pass a similar resolution, stating "that this Association is of the opinion that the vast majority of the population, which is purely native, and especially the trading and mercantile classes, are not aware of any cogent and convincing reason having been adduced for the proposed substitution of standard for solar time." As one newspaper succinctly summed it up, those benefiting from the reforms constituted but "a few hundred globe-trotters and exalted officials."[61]

A buoyant scene of associations of commercial as well as social and religious nature, a flourishing press in the vernacular languages of the subcontinent, clubs, and pamphlets—all attested to the growing importance of civil society and a public sphere in Bombay. In voicing their dissatisfaction with the new mean time, urban, British-educated Indian elites began to deploy these forms of publicity with increasing aptitude. Loudly asking questions about the representativeness of British circles could easily reveal other contradictory claims and practices by the colonial state. It was now time for the British Viceroy in India, Lord Nathaniel Curzon, to serve as a lightning rod for public anger.

Around 1905, a new and increasingly impatient generation of nationalists was growing disillusioned with the institutionalized form of anticolonial modernity that the Indian National Congress and similar institutions embodied. British-generated bodies such as the Congress, as well as the limited opportunities offered by municipal self-government, were now viewed as insufficient. Above all, it was Lord Curzon's Viceroyalty that pushed these more radical voices over the edge and led them to embark on a period of openly violent action against the colonial state, mainly in Bengal.[62] Curzon's first clash with Indian public opinion concerned education. In 1904, Curzon's government passed the Universities Act in order to widen government control over higher education, a step harshly criticized by leading early Indian nationalists. In July 1905, around the same month during which Indian Standard Time had been introduced, the much-despised Curzon announced what would come to be known as the first partition of Bengal: the eastern parts of the previously largest administrative unit in British India would be united with Assam, the west with Bihar and Orissa. After Curzon's plans for partition became known, members of the educated Bengali middle classes organized a protest movement that for several years successfully boycotted British goods and instead propagated the consumption of local, Bengali products. The so-called Swadeshi movement was later seen as a foil and precursor to Swaraj, Gandhi's call to self-rule. Partition was perceived as an outrageous disregard of public opinion and a cold-blooded exercise in "divide and rule."

In this atmosphere, opposition to the new time was quickly fused with critiques of British actions and the colonial state more broadly. One newspaper described uniform time as "adding fuel to the flame of intense dissatisfaction prevailing in the country."[63] Curzon's "style" of leadership was increasingly viewed as arrogant and manipulative. Once the Bombay Chamber of Commerce revised its previous stance toward time unification and suddenly voted in favor of it, newspapers were quick to point fingers at the government and Curzon, whose influence they suspected behind the sudden change of heart. The inhabitants of Bombay were reported to view the introduction of standard time as one of the "despotic measures thrust upon the Government of Bombay by the late Viceroy of India."[64] Another article mocked Curzon's presumed megalomania. "It would seem that there is nothing small or great to which the present Viceroy, with his indefatigable energy, would not apply himself," the piece stated. "Having performed his Herculean miracles in matters administrative and executive, . . . His Lordship . . . seems to be engaged in the task of over-riding Nature herself." Curzon was called an "imperial Caesar," and it was his "'craze' for uniformity" that led the viceroy to ordain the unification of time.[65]

When the government had passed certain measures to prevent the spread of bubonic plague, specific needs and conditions among various religious groups had been adduced as a justification for the uneven application of some of these policies. Aware of the contradiction and possibly also alluding to the partition of Bengal, a newspaper article exclaimed, "Here is the Indian government in one breath declaring that it is inexpedient to have uniformity in all important matters where large masses of people are concerned, and in another breath dwelling on the beauty of uniformity!!" And it added, "The plea of uniformity is worthless. It cannot hold water. Even so civilized and progressive a country as Europe [*sic*] has never adopted uniformity in the sense the Indian government has forced it on almost the whole of India save Calcutta."[66] When Balachandra Krishna raised his voice at the Madhav Baug meeting, he too chastised the recent obsession with uniformity, stating that "a movement had been going on for the last four or so months to introduce standard time in India. . . . There was a sort of mania for equalizing, unifying, and bringing into uniformity, several things. . . . Standard time was the outcome of that idea."[67]

The adoption of a mean time for British India occurred as gradually as similar stories played out in Europe and other parts of the world. Time remained variegated; multiple times coexisted. In the following years, with some regularity, initiatives sought to rein in Bombay's unruly times. In 1908, certain sections of the Municipal Corporation put standard time on the agenda again. Ever since 1906, municipal clocks had been following Bombay time, while the Government of Bombay continued to operate standard time throughout the presidency. Another protest against "this fresh attempt to insidiously force the hated time on the people" arose no sooner than the announcement had been made. The Indian population of Bombay, "the backbone of Bombay's commercial and industrial life," was expected to "muster strong and give out pretty vigorously its bit of mind" at the upcoming assembly. As of 1908, nobody was keeping standard time anyway "save the official classes and the officialized non-officials, the tribe of the parasites and flatterers," a newspaper scorned. "For Heaven's sake don't force it upon an unwilling population already breathing sullen discontent," another paper pleaded.[68]

Years later, in 1927, a proposition came before the Bombay Municipal Corporation to universalize standard time in all municipal offices and to convince the municipal government to give up the time it had adopted beginning in 1906. The move failed, and Bombay went on following two times. Voices could now be heard complaining that retaining Bombay time imposed hardship on whoever still followed it, for many institutions beyond municipal offices had gradually come to switch to standard time.[69] In 1928 and again in 1929, the Municipal Corporation was facing another vote on standard time. The ghost of Sir Pherozeshah Mehta still haunted Bombay politics. This time the opponents of standard time, by far outnumbering those in favor, made much of the fact that the now deceased Pherozeshah Mehta had opposed the time change and that the Government of Bombay had neglected to consult the municipality as the local self-governing body. For the sake of his memory alone, standard time should be opposed. And so it happened.[70]

After the split in 1906, standard time was discussed again in 1918, 1921, 1924, 1934, 1935, 1939, and 1942, to no avail. So principled and entrenched had the frontlines become that newspapers now spoke of the "Battle of Clocks" when writing about Bombay's time standoffs.[71] The arguments advanced for retaining Bombay time cited the religious duties of the city's various religious communities as tied to sun time, the historical legacy of Curzon's silly and offensive step, and the patriotic chants with which the Municipal Corporation's decision to reject standard time was greeted in Bombay. In the 1930s, most official institutions in the city had since changed their clocks to standard time; the Municipal Corporation was now the lone purveyor of a time that appeared out of character. The truculent adherence to Bombay time was ridiculed when commentators asked, "How much longer is Bombay going to be the village that voted the Earth was flat?"

If support for Bombay time was waning, it still took fifteen more years for standard time to triumph over the proud Municipal Corporation. In 1950, the "44-Year-Old Battle of Clocks" came to an end when, soberly and without much ado, clocks were set to Indian Standard Time.[72] It is not known to what extent other cities and regions across the subcontinent clung to local times or slowly adopted the new mean time of UTC+5:30. Calcutta never officially performed the switch to standard time in the first place. In 1919, a newspaper could still suggest that "Calcutta already possesses more times than she knows what to do with."[73] Madras may have kept local time as late as 1939 as well. Did the British government in India flinch from unifying time for so many decades out of an ambivalent attitude adopted toward governing India after the rising of 1857? Following the so-called Indian Mutiny, the East India Company's rule was ended and much of India came under direct control of the British crown. More conscious than the company overlords to avoid offending Indian religious habits and customary practices, the new British Raj promised to respect public opinion more than previously.[74] But the torrent of documentation and codification projects produced in the following decades and aimed at "conserving" indigenous customs and laws was always ambivalent and could serve the regulation and policing of Indians just as much as it purported to shield the local from outside influences.

A different set of reasons is more instructive for understanding colonial time politics. The meaning and function of mean times was a national or regional one in Europe and a territorial or regional one in colonial settings where Europeans slowly introduced such times. Mean times assumed the role of easing coordination in one colony, or perhaps at best in a regional cluster of possessions, but were not yet viewed as elements in a grid of worldwide time zones. British administrators thought of India as a vast, decentralized territory rather than a single territorial space. Keeping a variety of local and regional times thus did not seem all that extraordinary under such circumstances.

British India, then, was another beacon of variegated timescapes until the middle decades of the twentieth century. As in Europe, adopting mean times added layers of time. Living and working among plural times came naturally to the inhabitants of Bombay, a city where the plurality of times was enhanced by the coexistence of different European and Indian trading and religious communities. When authorities decreed the use of standard time in Bombay, they stoked resentments that had been brewing over other unpopular measures, above all university politics and the partition of Bengal. In the process, the strong foundations of local urban Bombay identities were mobilized to now fuel actions with a decisively national bent, criticizing colonial politics and verbalizing anti-British sentiments. The city thus came to stand for the nation. In situations where no nation-state existed, or, put differently, where the realm of the national was occupied by a colonial power, the universalizing reorganization of global time engendered alternative visions of identity when local and regional concerns like Bombay time and the partition of Bengal suddenly turned into national grievances.[75] In their objections to time unification, opponents of standard time clung to Bombay time and harnessed the motivational forces of their urban identity. Local matters became a valve for communicating disapproval of imperial rule, and the stated grievances were claimed to bear relevance for all of India. Insistence on local time positioned Bombay and India outside the British realm: by refusing to adhere to the time of the colonizer, early Indian nationalists insisted on the possibility of carving out autonomous timespaces that did not stand in any meaningful relationship to the prime meridian at Greenwich. These were instances where colonized societies refused to be mapped onto the hierarchical grid of universal historical times that reflected Euro-American norms of progress and modernity. Bombay and, by extension, Indian time were defined not by British time and supposedly universal history as the measure of all things, but out of India itself.

CHAPTER FIVE

# Comparing Time Management

In late Ottoman Beirut, a plurality of times was always within eyesight or earshot. The members of the several sectarian denominations centered in and around the Levant's main port counted time differently and called it differently. They carried different calendars around with them. They may have prayed to the same God but did so at different times, observing different rhythms and schedules, announced by church bells and the Muezzin, respectively. The Western missionaries who had descended upon Mediterranean shores to salvage their souls from doom brought with them yet another understanding of time, which they often equated with discipline. Members of the local merchant classes were but vaguely interested in the stipulations of the common God and his European and American mouthpieces alike, and preferred time saving over soul saving. And yet, what united these different constituents of Beirut times around 1900 was their pronounced interest in time—whether "time" was understood to signify ways of counting time, different calendars, clocks and watches, or the use and management of time.

The Ottoman provincial capital of Beirut represents a certain type of city found in the world of 1900. Often ports, often colonial cities, such urban centers embodied the global condition of the nineteenth century like few other spaces did. Trade and commerce lured in foreigners in search of business opportunities; a mélange of denominations attracted Euro-American missionaries and raised their hopes of wresting a few converts from the clasp of what clearly had to be inferior religions; colonial administrations brought foreign government officials to town. Due to their economic centrality, these global cities were often simultaneously centers in the production and dissemination of print. As a consequence, local religious leaders and members of the new middle-class professions, ever keener on availing themselves of instruments for cheaply and rapidly communicating their ideas, flocked to these sites as well.

Intellectual, economic, and political hubs, such cities became incubators of modern globality and, as part of this condition, temporal pluralism. Euro-American concepts and ideas often passed through these cities and their intellectual leaders and printing presses when moving to the non-Western world. It was here that local thinkers translated and adapted some of these ideas to resonate with the intellectual and religious traditions of their own societies. In other instances, the presence of Westerners, their habits and ideas, aroused the criticism and ire of those encroached upon by a growing foreign presence. In the world of the long nineteenth century, Alexandria, Buenos Aires, Cape Town, Dakar, Hong Kong, Istanbul, Kolkata, Mumbai, Odessa, Shanghai, Tangier, Thessaloniki, and Trieste, among others, can be counted among such cities.[1] In these cities, globalization took on a very local guise.

The inhabitants of a city like Beirut did not experience time changes primarily at the hands of a colonial or indigenous state and its retinue. As in so many other parts of the world, the introduction of a region-wide mean time came late in the Levant and was loftily applied as late as the 1930s. The state here was an uncertain source of time, but this did not mean that the global preoccupation with time and "time talk" were absent from such locales. Contrary to Euro-American perceptions of the "East" as time-indolent and stagnant, Beirutis vividly engaged with different expressions of time in these years—public time, private time, religious time, clock time, calendar time. In the multitemporal environment of global cities, non-Western societies proved less oblivious about time than Euro-American observers portrayed them.

In the late Ottoman Levant, a preoccupation with time management and the use of time mirrored the British and European obsession about "wasting" daylight and thus surrendering potential time for useful activities. In the Middle East as a whole, a more interconnected world could be viewed to bear threatening features. Since the middle decades of the nineteenth century, local commentators looked on in dismay as the Ottoman Empire was struggling to reform itself in the face of European pressures. In this situation, local journalists and intellectuals discovered time as a crucial feature of their own as well as Euro-American life. In the writings of Arab authors, time management and a crusade against wasting time became instruments for the self-strengthening of the Eastern, Arab, and Islamic civilization. But in the course of working through concepts of time management, Arab intellectuals transformed time on and in their own terms. The circulation of ideas and the globalization of time produced a "nationally" interpreted, civilizational Arab and Islamic time.

Local conditions fostered an apprehension of temporal pluralism. Living and speaking multiple times came naturally to the denizens of nineteenth-century global cities. Much more readily than their European counterparts, the writers and readers of the region's many periodicals made a habit of comparing times—how to compute times, how to convert them, how to call them, and how to use them wisely. For these comparisons, Arab contemporaries relied not least on a lively publishing scene of almanacs and calendars, timekeeping devices that visualized the comparability but also incommensurability of times in unique ways. When it came to imagining time in more than just one of its varieties, Arab contemporaries were not only less oblivious about time than Europeans and Americans claimed, they were in fact more apt than many Europeans at picturing varieties of time.

• • • •

Beirut's preoccupation with time was framed by several developments both internal and external to the historical region referred to as Bilad al-Sham or Greater Syria.[2] Over the course of the nineteenth century, Beirut grew from little more than a fishing hamlet of roughly 8,000 inhabitants to a bustling port city of more than 100,000. By 1900, Beirut had emerged as the uncontested port city in the Levant. Beirut's rise was propelled by its growing integration into the world economy. Around mid-century, European commercial expansion into the silk industry of the mountains surrounding Beirut moved the city into European political and economic orbits of influence. Soon, Europeans and Americans opened a slew of consulates and businesses, and later, schools and their own extraterritorial courts of law.[3]

At the same time, the middle decades of the nineteenth century saw a redrawing of provincial boundaries that paved the way for incorporating provincial towns like Beirut more closely into the new "geo-administrative hierarchy of centralized Ottoman rule."[4] Exposure to the world economy unsettled the economic and social balance that had persisted in the region. The Ottoman province of Beirut was made up of a fine-grained fabric of multiple religious denominations sharing the cities of the coastal strip and the mountainous hinterlands. In 1860, sectarian tensions, often fueled by dislodged social and economic fortunes, erupted into a civil war that was fought out mostly in the mountains between Christians and Druze. After an estimated 20,000 people had been left dead, the conflict ended with an international "humanitarian" intervention by European powers on behalf of the Christian population. After the civil war of 1860, Beirut and Damascene notables engaged in a tug of war over the reorganization of administrative boundaries following from the conflict. In both locales, lobbyists petitioned the Ottoman government to upgrade their city's administrative status at the expense of the other side. It took more than two decades and several setbacks, but in 1888 Beirut came out on top in this regional power struggle as the sultan granted the creation of a province for Beirut.[5] As the new seat of power and commerce, Beirut soon became the site of a veritable construction frenzy in transport and communication. In 1860, Beirut became the first city in the Bilad al-Sham to open a telegraph station. Three years later, a major road between Damascus and Beirut was completed. European steamers now regularly passed through the port of Beirut. As of 1895, a railway line connected Beirut and Damascus. Tramways began circling the city in 1907.[6]

Several religious denominations cohabited in the city's urban space. Beirut was host to (mostly) Sunni Muslims; Greek Orthodox, Roman Catholic, and Maronite Christians; and a small Druze population of fishermen on the Western shore of the city. The provincial capital of Beirut was a "polyrhythmic" city in which religious rituals and the schedules of modern technology alike inscribed time onto the lived experience of the urban environment.[7] Different temporal rhythms came with their own nomenclature in tow. By the late nineteenth century, two systems of counting hours existed in parallel in certain parts of the Ottoman Empire. One was the common European method, referred to as "Frankish" in Arabic or "alafranga" in Ottoman Turkish. The other still widespread system of computing time was termed "Arabic time" or "alaturka/Turkish time." In the Islamic tradition, the day began at sunset, and sunset was reckoned as 12:00 noon. Hours were counted from sunset in two cycles of twelve hours each. As in Europe, these hours had for a long time been uneven or "seasonal" hours. With the spread of mechanical clocks, uneven hours mostly fell out of use. If hours were now even, watches still had to be adjusted every day, for the time of sunset varied ever so slightly with the seasons. *Alaturka* or Arabic time abided until the first decades of the twentieth century when it was gradually replaced with European computations of time. When printing steamboat or tramway schedules and other indications of clock time, newspapers, guides, and other publications therefore routinely added a specification of "Frankish time" or "Arabic time."[8]

Beirut's temporal pluralism was manifest in the city's built environment. The fivefold daily prayer structured the lives of the Muslim population, announced forcefully by the call of the Muezzin from atop a minaret. For Christians, there was Beirut's Anglican parish church, which had received a "fine bell" and tower clock from a church in New York City.

The Maronite cathedral, another church, donned not just one but two bell towers.[9] Another temporal landmark originated with one of the newly founded missionary institutions in the city. In 1866, American Presbyterian missionaries opened the Syrian Protestant College (SPC; renamed American University of Beirut in 1922). As a vivid expression of the occasionally fierce rivalry between Catholic and Protestant missionaries, French Jesuits moved their seminary from the small mountain town of Ghazir to Beirut in 1874 and soon thereafter acquired certification from the French government as a full-fledged university (Université Saint-Joseph des Jésuits). The Jesuit university installed a smaller clock on its main building.[10]

In the contest over public time at least, American Protestants won the day. College Hall, the SPC's main building, was embellished with a clock tower and powerful bell, visible and audible far beyond the confines of the new campus. Henry Jessup, an American missionary, could barely conceal his exuberance: "The citizens of Beirut, Moslems, Christians, and Jews, were so anxious to see and hear a clock whose striking could be heard throughout the city, that a local subscription was raised. . . . Thus the Mohammedans who abominate bells, and the Jews who dislike Christian churches, contributed to the erection of a Christian bell-tower. And when the clock was finally in place and began to strike the hours, crowds of people gathered in the streets to hear the marvelous sound."[11]

The tower clock at the Syrian Protestant College was to serve the institution internally as well, once students began to arrive on campus. During the early 1870s, when the first buildings were under construction, the future president of the college, Howard Bliss, worried, "I do not see how we are to keep the exact time at the College. Our old clock runs fast and slow without order. And the bell too is not forthcoming. . . . The faculty and students are anticipating a military and iron rule this coming year out at the new building."[12] The College Hall clock received its time from the astronomers at Lee Observatory, opened as part of the Syrian Protestant College in 1873. Lee Observatory also furnished the city of Beirut with a time ball service. As in so many port cities around the globe in this era, "for the purpose of supplying accurate time to the residents of Beirut and to visiting navigators," a time ball was raised daily at 7:55 a.m. and dropped at 8:00 a.m. local time. A former director of Lee Observatory later recounted that people living in the nearby hills in places such as Broummana bought small telescopes to watch the ball drop. As late as the 1930s, the timekeeper who was responsible for timing the Islamic prayer at Beirut's main mosque (al-ʿUmari) reportedly relied on the clock tower in timing the call for prayer.[13]

Beirut's most recent time monument was not the purview of any religious denomination but that of the Ottoman government. In one installment of a multipronged reform program, Sultan Abdülhamit II sought to represent the Ottoman state and the center in Istanbul as a benevolent protector and, at the same time, modernizer of the remote provinces.[14] Ceremonies and rituals were as much part of such gestures as architectural representations. When the sultan granted permission to construct clock towers in several provincial cities of the empire, this certainly carried all trademarks of a move to bestow the temporal modernity of the center onto the provinces. Many cities in the Ottoman Empire erected clock towers, some significantly earlier than Beirut, others as part of a construction frenzy surrounding the twenty-fifth anniversary of Abdülhamit's ascension to the sultanate. Yet in the case of Beirut's new clock tower, it was the confluence of local initiatives by the Beirut provincial council with imperial policies that led to the construction project.[15] In 1897, the Ottoman Governor General of the province of Beirut addressed the sultan on behalf of local interests. He started by outlining the several "foreign institutions" that had established "clock towers with bells" in the city, "all of them with a western clock. Because there is no public clock which shows the mandatory Muslim (prayer) times Muslims, even officials and (other) civil servants have regrettably had to adapt to the time of foreign clocks." There was, therefore, an "urgent need" for a public clock that would determine the religious times of Muslims.[16]

Once the project was approved, Yussuf Aftimus was charged with overseeing the construction of the clock tower. Aftimus was a graduate of SPC and current municipal engineer who had worked briefly for the Pennsylvania Railway Company and General Electric after finishing his graduate studies at Union College in New York. Aftimus subsequently designed the Ottoman and Persian pavilions at the 1893 Chicago World's Fair and was widely considered an expert in Ottoman architectural style. Beirut's new clock tower was situated prominently in front of the main Ottoman administrative and military buildings in the city. Like Aftimus himself, the Ottoman clock tower epitomized a blend of styles and Western and Eastern traditions: two of the four clock dials on each of the tower's four sides used Latin numerals, the other two Arabic ones. Beirut's temporal pluralism may have been exceptional even by the standard of nineteenth-century global cities. But the medley of religions and ethnicities in many of these localities suggests that a similar element of diversity existed elsewhere. At these sites, temporal pluralism was a daily experience.[17]

• • • •

BY THE EARLY DECADES of the twentieth century, Beirut was not yet facing an Ottoman or European colonial state forging a new mean time. Ottoman officials were examining various questions pertaining to time, but for the moment, none of these measures came to pass. The Ottoman government in Istanbul was discussing the replacement of "Turkish" time with European times in the context of the Young Turk Revolution of 1908. Briefly after this attempt to steer the Ottoman Empire in a more constitutional direction, a committee considered the abolition of the Ottoman fiscal calendar (a special calendar year beginning in March, used in calculating salaries and for other bureaucratic purposes) and the adoption of European time, but the cabinet punted.

Behind these efforts was Gazi Ahmed Muhtar Pasha, formerly a senior Ottoman officer in Egypt who taught mathematics and astronomy and authored books on timekeeping and the calendar.[18] In 1910, Muhtar Pasha returned to Istanbul, where he brought before the upper house of the Ottoman parliament a bill that proposed to make mandatory the display of clock dials showing European mean time next to *alaturka* ones on all public clocks as well as on clocks used at mosques, but the bill did not go far. In 1912, the Ottoman state eventually adopted European time in the military and in bureaus of the civil service. Following the international time conferences in Paris in 1912 and 1913, the Ottoman radiotelegraphy station in Istanbul received a daily time signal from Paris. Even at the center of power in Istanbul, however, the spread of European time remained patchy at least until the 1920s.[19]

As far as it could be established, Greenwich time was introduced to Beirut in 1917 with the arrival of French troops and probably formally institutionalized as part of French mandate rule a few years later.[20] But as attested by almanacs and calendars published in the 1920s and even 1930s, the actual implementation of the time of UTC + 2 must have remained superficial. Such publications continued to explain what mean time was in the first place and what Beirut's time zone was. Almanacs and calendars moreover listed the local times in a range of bigger cities (including Beirut) around the world at the time of 12:00 noon at Greenwich or sometimes Paris. As elsewhere, the introduction of mean time in Beirut merely added yet another layer of time without suppressing religious and secular local times.[21]

• • • •

WITH NO IMMEDIATE official move to introduce mean time to the city of Beirut, Levantine newspaper readers and authors nevertheless paid keen attention to time. In the decades around 1900, Beirutis showed a growing preoccupation with clocks and watches. This uptick in interest was well registered in European commercial publications reporting on business opportunities in the Levant. "The demand for cheap watches is very considerable, and Beyrout [*sic*] is the principal center from which the interior as far as Medina [Saudi Arabia] and the coast towns are supplied," one such publication wrote.[22] The report further specified that the best-selling watch, "met with everywhere," was the Swiss-made Système Roskoff, available with either Arabic or Latin numerals. The German (later turned Swiss citizen) watchmaker Georges Frederic Roskopf (1813–1899) made a name for himself as a manufacturer of affordable watches. In the 1860s, he offered a watch that was being sold at the price of an ordinary worker's weekly loan, a timepiece he called the "proletarian watch." Such timepieces aided the spread of watches among workers in Europe and also rendered them more affordable and attractive in the non-Western world.[23]

Concurrently, a growing number of Beirut watchmakers were now listed in local guide books and almanacs among the city's businesses. The trend toward a production of low-price mass watches notwithstanding, such devices likely remained luxury items, with prices beyond the reach of the general public. Watch merchants and watch repair shops were commonly located in the area of Beirut's Souk where sellers specialized in some of the more luxurious goods offered in the city.[24] Yet judged by the appearance of watch advertisements in Arabic newspapers of the period, there must have been a small but sizable enough market to support those peddling in clocks and watches. Merchants often praised their services by claiming international experience. The representative for the Swiss Longines watches in Syria offered timepieces from the "biggest factories in Europe." One Jirji Bishara Hajji had spent five years in Australia as the representative for the American Waltham company and upon returning to his Levantine homeland "brought with him all the modern machines used in this industry." A nineteenth-century encyclopedia of crafts and professions practiced in the city of Damascus even deemed it necessary to warn readers of the black sheep that had infiltrated the profession without proper knowledge and training, unjustifiably claiming the title of *sāʿātī* (watchmaker), so popular had watches become.[25]

When not selling or buying watches, the reading public of the Levant was interested in the timepieces of others. Descriptions of famed historical clocks or prominent monumental ones featured frequently in the local press. Around the time when the Ottoman clock tower was discussed and prepared, an article described "The Clocks of London." The city was furnished with "many big clocks raised on high towers" that "made it easy for many people to know the time when needed." Such articles also reported with awe on newly installed systems for controlling the distribution of time via networks of clocks in large European cities, whether through pneumatic installations as in Paris or by electricity as in Vienna and Berlin. Historical and particularly artful and sophisticated clocks were another frequent object of admiration. The English King Henry IV kept a clock in his bedroom that would run for one year without requiring any winding, papers marveled.[26]

Another article described a clock manufactured by the American watchmaker Waterbury of Connecticut, a masterpiece that had taken twelve years to build. As part of the clock apparatus, an installation depicted "the progress of the current generation and man in science and industry," including scenes depicting cotton cultivation and coal mining, sowing machinery, textile manufacturing, electric appliances, the telegraph and the telephone, as well as comparisons between old, outdated watch technologies and the latest innovations of the trade. What late nineteenth- and early twentieth-century Arab journalists and essayists admired most about historical clocks was the skill and craft that went into manufacturing them, and clearly, their presumed accuracy and longevity. Precision was becoming a coveted value.[27]

The mounting Levantine obsession with clocks and watches was immortalized in the form of a short story. Its author, Mikha'il Nu'ayma, immigrated to the United States at the beginning of the twentieth century, one of many Syro-Lebanese who left behind the declining fortunes of the silk industry in the Lebanese mountains for a new life in the Americas or Australia. In 1929, he wrote "The Cuckoo Clock," the story of a peasant whose fiancée is lured away from a mountain village by an emigrant who had temporarily returned to his homeland.[28] Here, the temporary returnee showed off several flashy gadgets of modernity and, in particular, a cuckoo clock he acquired in America. The speaking bird utterly fascinates the village community. After his fiancée disappears across the ocean, the peasant draws a balance of his life so far, "and for the first time in his life he detested everything his eyes saw as hideous and disgraceful, his oxen and plough, his trees and vineyards."[29] The peasant then decides to try his luck in the country to which his fiancée was whisked away with the help of the clock and becomes a millionaire in America. Meanwhile, his former fiancée is a wrecked woman who has to earn a living working in a bar, for the clock-owner abandoned her long ago. Filled with disgust at the oversaturation of civilization in the New World, the peasant-turned-millionaire eventually returns to his native village in the Lebanese mountains where he grows old preaching the beauty of nature and the pleasure of the simple things in life.[30]

A growing availability of clocks and watches required Levantine audiences to develop a language of time and timekeeping. An Arabic-French language guide circulating in the Levant featured a section on expressions related to time, including a hypothetical visit to a watchmaker. The French phrases for which Arabic translations were offered in the language guide included "At what time is sunset?," "How quickly time passes," "I kill time sometimes by reading or by taking a stroll," "The Azan struck noon" (the Islamic call for prayer), "I do not know. My watch is not accurate," "I did not wind my clock," "I did not adjust my clock," "It stopped," "It is still not working," "It is slow," "It is fast," "My watch is old and needs repair," "Do you have watches from Geneva?," "I need it as soon as possible since I can't be without a watch," and "My work forces me to carry a watch every day."[31] These phrases captured the increasing dependence on and familiarity with timepieces.

As timepieces grew more popular and prolific, a different kind of timekeeper spread alongside clocks and watches. Talking about different aspects of time first and foremost operated within a comparative framework. Arab authors writing in the flourishing press were consumed with comparing themselves and their time management to Europeans in order to gauge how Arabs stacked up and to identify what to change if Easterners wanted to dodge European colonialism. Europeans and Americans, on the other hand, could not help but notice the stark differences between their supposedly regular and uniform method of keeping time and whatever local societies practiced. More generally, evolutionary thought, whether of the Darwinian or the more social, Spencerian variant, provided all parties with a template for thinking in comparative stages of civilizational development. Comparison was an important epistemological tool of the nineteenth-century world for placing oneself in that world and for relating to other societies. It was an intellectual operation central to the formation of a global consciousness.

The most illustrative visual expression of the comparative mind was a tool not commonly counted among timekeeping devices. In the world of the Eastern Mediterranean, these markers of time may have easily proved more influential than clocks and watches due to a much lower price. Almanacs and calendars on sale in Beirut and the Eastern Mediterranean were increasingly sought-after from the 1880s. Beirut with its numerous religious denominations was a city where calendar pluralism was taken to the extreme. At least four calendars formed part of everyday life: the reformed Gregorian calendar; the unreformed, Julian calendar used by various churches of the East; the Islamic lunar Hijri calendar; and the Ottoman "Rumi" or sometimes financial/"Maliyye" calendar. The latter used the Hijri era for counting years but Julian months, with the year starting on March 1. The Ottoman calendar was an administrative device above all other things but was used in many official announcements as well. All calendars in use came with their own holidays. In the latter decades of the nineteenth century, printing presses put out not only books, journals, and newspapers, but also calendars and almanacs designed to help navigate the calendar pluralism of the Levant and Egypt. As a rule of thumb, every printing press and major newspaper in Beirut and Cairo seemed to have published an annual almanac or calendar beginning in the 1880s, and many independent and specialized publications (e.g., agricultural calendars) added to the lot.

Newspapers and journals contained a growing number of advertisements announcing the arrival of the new "al-Hilal Calendar" for the coming year, providing brief descriptions of contents and price. Shortly, almanacs would feature texts, mostly pieces on recent technological inventions like the telephone, health advice, and science news. These print products functioned as extensions of the flourishing press but have never systematically been studied or used for historical research. On their pages, calendars and almanacs commonly listed a full annual calendar with Hijri, Gregorian, and even Jewish and Coptic weekday names, month names, and years. Indications of sunrise and sunset in both Arabic and Frankish time completed the calendar information. Other sections of these almanacs named Ottoman government officials in the city and foreign consuls and printed steamboat and tramway schedules. Depending on the religious background of the editors and the press, some almanacs prioritized information according to their own denomination's needs, while others were more deliberately multireligious. On the pages of almanacs and calendars, times could literally be experienced and compared side by side. Arguably it was this visualization of the coexistence of multiple calendar times that made comparing different calendars a popular pastime.

Many of the periodicals produced between Beirut, Cairo, and Alexandria dedicated a certain amount of space to Q & A sections. Readers wrote to the editors with their questions, and editors set out to answer these requests for clarification in one of the following issues. Since the turn of the century, readers asked for advice about comparing calendars, about converting one date into another, and about explanations for differences between calendars. One Mursi Sadiq writing to *al-Hilal* begged to know the explanation for the twelve-day difference between the Frankish (Gregorian) and the Julian calendar and what historical developments had led to the divergence. A different article was devoted to the historical background of the 1582 reform of the Julian calendar. Several pieces talked about the history of different calendars in use among the ancient Egyptians, Babylonians, Greeks, and other societies of the ancient world. Intercalation, and why it was necessary, was another favorite subject of inquiry. To a greater extent than their fellow Euro-Americans, Arab readers lived in multiple times. While letter writers may have demanded clarification primarily on the historical background of the Gregorian reform, they easily comprehended a variety of times to be a natural state of affairs. To them, a world of one immutable time would have appeared odd. In the view of Arab authors and readers, globalization and interconnectedness were more about contriving instruments and methods to compare and convert different calendars than about replacing difference with homogeneity altogether.[32]

Numerous books and pamphlets of the second half of the nineteenth century discussed the Islamic lunar calendar, the history of calendars, and time and date conversion. In several publications, Mahmud Pasha "al-Falaki" ("the astronomer"), one of Egypt's most famous astronomers of the nineteenth century, weighed in with books and essays on calendar comparison and conversion charts, the history of Arab calendars prior to the advent of Islam, and the astronomical secret of the pyramids.[33] Gazi Ahmed Muhtar Pasha, the Ottoman official who brought the proposal to use European time before the Ottoman parliament, penned similar texts.[34]

The availability of calendars and watches and the presence of clock time in public space gave new meaning to accurate and precise time. Over several weeks in 1892, the veracity of published information on time came under scrutiny from local newspaper readers. The paper *Thamarat al-Funun,* considered the main outlet for Muslims, got in a spat with *al-Bashir,* one of the papers published by the Jesuits. The object of contestation was a time of sunset that *al-Bashir* had printed in the almanac it published every year. Readers had written to *Thamarat al-Funun* asking for clarification, as the indicated times appeared wrong or at least questionable to them. And *Thamarat al-Funun* felt obliged to pick up the thread since, as it explained, Muslim prayer and fasting rituals depended on some of these times.[35] The confusion arose over the time of sunrise given for January 10, 1892, as 2:12 Arabic time (roughly 6:30 a.m. European time), a time that to *Thamarat al-Funun* and its readers appeared to fall too early. *Thamarat al-Funun* also arrived at the conclusion that 2:12 Arabic time was the latest the sun would rise on any day of the year according to *al-Bashir*'s calendar. Based on this finding, *Thamarat* accused *al-Bashir* of making January 10 the shortest day of the year, while everybody knew the shortest day to fall at the winter solstice, December 21.[36]

*Al-Bashir* felt the need to clarify its position by first explaining the difference between counting Arabic and Frankish time, to which *Thamarat* coolly replied, "The terminology of the Europeans does not concern us." *Thamarat* and its readers had contended that a sunrise time of 2:12 was "contrary to what is perceived," thus contrasting the power of human eyesight and objectivity with presumably false calculations. *Al-Bashir* found a way to clarify the presumed mistake: its calendar was laid out not only for the city of Beirut but other towns in the province and even parts of the province of Syria as well. The time of 2:12 was not meant to indicate sunrise in the city of Beirut because the mountains of Jabal Lubnan (east of Beirut) occluded the vision and obstructed the appearance of the sun by an average of twenty minutes after it rose above the horizon. "And no calendar is to blame for that," *al-Bashir* declared.[37] Out of precaution, *al-Bashir* had therefore dropped between eight and fifteen minutes from the time of actual sunset to account for the elevations an observer would encounter when standing somewhere in the mountains of Jabal Lubnan east of Beirut.[38] The matter remained unresolved; it suggested a growing interest in accurate and truthful indications of time.

● ● ● ●

THE MOST PERVASIVE ENGAGEMENT with time by local journalists, intellectuals, and publicists occurred in the context of Arab and Islamic self-improvement beginning in the second half of the nineteenth century. The Ottoman Empire faced an uncertain future in an age when multiethnic fabrics of the Ottoman sort threatened to be torn asunder by the centrifugal force of nationalism. Self-strengthening, among other aspects of time management, would brighten up prospects in the age of imperialism. The Ottoman Empire had been losing territory since the beginning of the nineteenth century and conjured up the famous image of the "sick man on the Bosporus." The sultan sought to counter any such perceived and real threat of disintegration by embarking on a vast reform program around the midcentury, collectively referred to as the so-called *Tanzimat* (reorderings or reorganizations).[39]

In the peripheral provinces of the empire, the sultan's reformers were instructed to tighten and streamline administrative processes. In some cases, as in Egypt, such efforts proved too late. Here, a skilled military leader, who had been dispatched by the sultan to wrest back Egypt from Napoleon, installed himself and his successors in the position of a viceroy (Khedive), nominally a vassal to the Ottoman Empire but de facto an increasingly independent Egyptian ruler. Mehmet Ali, as was his name, embarked on a set of ambitious technocratic reforms of the military, taxation, and education in particular. Yet reforms had to be paid for, and Egyptian leaders turned to foreign investors to finance their visions of technocratic modernity. When a nationalist uprising threatened Khedival rule and bondholders alike, Britain intervened and occupied Egypt from 1882.[40] At this point, Arab observers in other parts of the Ottoman Empire sensed alarm. In cities such as Damascus and Beirut, a growing number of intellectuals, journalists, writers, educators, and other members of an emerging middle class perceived an acute twofold, Ottoman and European, challenge. Ottoman weakness and imperialist pressures drove home the need for the improvement of "Easterners" in the face of European threats. Otherwise, as the colonization of Algeria, British India, and now possibly Egypt suggested, Arabs and Muslims stood doomed to a similar fate of alien rule and subjugation. The beginning of the "scramble for Africa" in the 1880s reinforced such general fears.[41]

The intensification of imperialist competition and the occupation of Egypt jolted contemporary observers to question business as usual. Their answer to a sobering outlook was to encourage self-reform and self-improvement by "reawakening" fellow "Easterners" to embrace the best of their own history and by prodding them to adopt the most promising aspects of European science, industry, and culture. The resulting so-called *Nahḍa* (Arabic for renaissance, revival) was a reform movement in the broadest sense.[42] Its most outstanding feature was its eclecticism, an amalgamation of intellectual production and adoptions of European thought. Most importantly, the *Nahḍa* saw the emergence of a flourishing scene of newspapers and magazines published in places like Beirut, Cairo, and Alexandria. Without these new channels of communication, translation, and dissemination, the frequent exchanges about time and time management would have been impossible.

Around 1850, an estimated twenty-four journals were published in Beirut. By 1875, eleven printing presses existed, publishing papers, journals, books, and almanacs. One of the most active early scholar-journalist-editors was Butrus al-Bustani, a Maronite Christian from the Chouf mountains southeast of Beirut who had converted to Protestantism. Al-Bustani was versed in Syriac, Latin, Hebrew, Aramaic, and Greek, published an encyclopedia as well as a modern Arabic dictionary, and authored a popular Arabic translation of the Bible. Upon moving to Beirut, he came across the work of the Protestant missionaries and for the rest of his life remained in close contact with the Americans as a teacher and translator. Starting in 1870, al-Bustani also published one of the earliest journals that offered a platform to reformist or *nahḍawi* thought, *al-Jinan* (Gardens).[43]

The different missionary enterprises populating the region continued their competition for the hearts and minds of the locals in publishing. Alarmed by the energetic efforts of Protestants like al-Bustani, Catholic Jesuit missionaries retorted with the weekly paper *al-Bashir* ("The Herald"), launched in 1870. In 1877, the Christian Khalil Sarkis started another semiweekly and later daily, *Lisan al-Hal* ("Voice of the Present"). Sarkis was another prolific figure of the literary revival. Together with al-Bustani, he founded a publishing house and later a press that specialized in the publication of Arab-Islamic classics. Author of at least nine books, Sarkis wrote about manners and ethics as much as school curricula and cooking. The first Muslim-run paper was spearheaded by ʿAbd al-Qadir al-Qabbani. Beginning in 1875, al-Qabbani published *Thamarat al-Funun* ("Fruits of Knowledge").[44]

One of the most lasting and influential *Nahḍa* publishing enterprises was *al-Muqtataf* ("The Digest"), founded in 1876 by Faris Nimr and Yaʿqub Sarruf at the Syrian Protestant College, where both had previously been students and now taught as instructors. In 1884, the editors moved the journal to Cairo after a number of public controversies over curricula at the SPC had disillusioned them toward the leadership of the college. An important later addition to the publishing landscape was *al-Hilal* ("The Crescent"), founded by the Syro-Lebanese Christian émigré Jurji Zaydan in Cairo in 1892. Zaydan the Christian was a vocal popularizer of Arab and Islamic heritage and history alongside scientific and other topics.[45]

Many of these papers were initiated and edited by Christians. But to understand nineteenth-century Arab print culture (and by extension, a preoccupation with time) to be Christian would be to miss the joint nature and variegated circulation of these publishing endeavors. Neither intellectually nor physically were Christians and Muslims separate epistemic and political communities in the late Ottoman Empire and Egypt. Publishers, writers, and their outlets were tied together by networks of print, Masonic lodges, societies for the advancement of sciences, and academies on the shores of the Eastern Mediterranean. Their editors frequently moved back and forth between Beirut on the one hand and Cairo on the other. The Arab press was enmeshed in networks that were transregional and cross-confessional and stretched from Beirut and Damascus to Cairo and Alexandria as well as many smaller cities in the Levant.[46] The circulation of Arabic newspapers and journals was often limited and rarely surpassed several thousand. But reading habits took on a form distinctly different from modern-day practices, and numbers therefore tell a partial story at best. Audiences in Beirut and especially of smaller towns in the countryside and the mountains accessed print products in libraries, schools, and government offices, often rendering reading into a collective act when newspapers were discussed on village squares, in cafes, and in homes.[47]

• • • •

ARABIC NEWSPAPERS usually contained summaries of political events in Europe (often either excerpted from European newspapers or later based on the digests and services offered by news agencies); reports from Istanbul about the sultan and Ottoman politics; translations from European scientific journals; literary pieces and segments of a serialized Arabic novel; local news; and a host of articles on a wide range of topics among which science ranked prime. By the closing decades of the nineteenth century, Arabic papers started carrying essays on the nature of man's time and life and ways of spending and using time. Arab authors, Muslims and Christians alike, looked around to compare the many times displayed throughout the region. In the gloom of an uncertain future, they noticed shortcomings, not the lively engagements with time that suffused Beirut's everyday life. A frequent complaint articulated in essays and articles was the tendency of "Easterners" to "kill time" by idling in cafes, oblivious of the true value of time. Such writings introduced Arab-speaking audiences to the much older notion of "time is money" and similar exhortations to save time by being productive instead of wasting it.

In 1909, a journalist named Jubran Massuh, writing for the newspaper *Lisan al-Hal,* penned an article titled "Dangers: How We Waste Time." One evening, he reported, he felt inclined to take a stroll on one of the promenades to "kill time" and to "pretend to be European." Massuh used the neologism "tafarnaja," derived from the word *ifranjī,* meaning Frank/Frankish, that is, European. To "pretend to be European" was a common, at times even mocked contemporary expression. He donned a tie and other European apparel and began walking without looking left or right, in conformity with the manners of Europeans. Until suddenly it dawned on Jubran that "Europeans do not waste time uselessly," for as they say, "time is money," and he quickly decided to return to his desk to "busy myself with something useful," saving the time he was about to waste.[48]

Prior to reaching his home, our journalist passed by "one of the important houses," possibly the domicile of influential notables, local elites who had traditionally been close to political power. The journalist heard clamor and noise emanating from the house and stopped to find that "twenty of the honorable men" had gathered. At first, he thought the dignitaries met in pursuance of "a noble cause, for a political conference, or an inquiry into an important and useful topic." Jubran Massuh waited to see if his assumption was right, and over the course of this interlude, witnessed nothing resembling what he had expected. One group was busy smoking; another engaged in trivial talk; others shared their admiration for a certain female object of their affection; two young men in a corner talked about "what the pen is too embarrassed to note." The first hour passed with such activities and gave way to a second, and the journalist still had not found out what the exact purpose of the assembly was. "The wine was already narrowing the minds of the people drinking it, confining them to a place beyond sound reasoning," Massuh deplored. After dinner, the group once more retreated for a smoke and, after a total of four hours, finally dissolved.[49]

Massuh then pitied his society for "killing precious time and wasting it to no purpose with useless issues." *Lisan al-Hal*'s author was quick to denounce himself for wasting four hours watching the dishonorable spectacle and for "being a partner in this disgrace," for conspiring in killing time. Massuh felt sadness upon looking at himself when he "returned to [his] Ottomanness," wishing he could be like the Europeans who did not waste time uselessly. Europeans progressed, advanced, and became civilized, improved the situation of their homeland, and strengthened their civilization by saving time and considering it to be money instead of wasting it. If Easterners would selectively adopt such secrets of European success, it would strengthen the Arab and Islamic civilization and permit Easterners to avoid full-fledged colonization.[50]

Time management as a tool for self-improvement was associated with a rigid system of dividing time into segments and dedicating single chunks of time to different activities. An article on "meal times" elucidates this widespread belief in scheduling and dividing up time. "Timing is very useful in a man's affairs and social life, just like a system and a good structure are of concern to different peoples and civilizations." Such structures and systems ought to hold a special place in the organization of meal times. It required a "choice of appropriate times" for meals to cultivate a healthy, relaxed body and a relaxed mind. This "organization of time" only occurred in "countries dominated by civilization," the article explained. A comparison of mealtime habits made this abundantly clear. One exemplary place where the organization of time was held in high esteem was England, where a sophisticated "system of mealtimes" was paired with "the useful English way to conduct work." The Englishman rose from his bed every day to eat a hearty and nutritious breakfast, "such as two or three lamp chops," eggs, pork, and potatoes.[51]

This practice stood in contrast with the French habit of "limiting oneself in the morning to tea and coffee" or "eating a little bit of jelly and sweets with a piece of bread," a breakfast pattern which in turn had been adopted "in the civilized countries of the East," to which the author likely counted his own home somewhere in the Eastern Mediterranean. The British worker, on the other hand, supported by a comprehensive and nutritious meal, could go on working into the late afternoon hours without requiring a full additional meal in the meantime. His breakfast gave him the power to last through half a day, and at that point, he only interrupted his work briefly to "eat a piece of meat in a thin slice of bread, the 'sandwich.'"[52]

The British worker "does not waste his time and interrupt the chain of his work during the earliest times of his day, because time is money," the article stated. The French on the other hand "waste their time and do not complete their work in one installment." It was French practice to start work around nine in the morning followed by a two-hour break to eat lunch, after which work was continued into the early evening hours, "something in which there is no usefulness, . . . even the opposite of usefulness," the article found. As for the author's own society, he concluded, "if we look at the best of what the leading civilizations have recuperated from all nations, . . . we will improve in the way of production, profit, and gain."[53]

A similar point about the division of time was made in a different article. Its author recalled his school days. Students working on numerous subjects at the same time without designating certain times for the study of individual subject matters normally lagged behind those who devoted themselves exclusively to one topic. If time was not divided and assigned in such ways, "someone who had set aside time for writing suddenly busied himself with reading, . . . and all activities are ruined in an instant."[54] At first blush, such a text might appear to be nothing short of an advertisement for improving productivity in line with the demands of industrial capitalism. Yet the motive behind encouraging regularity and a systematic division of time into segments was different. Economic gain was not the primary goal in the end but rather, the self-improvement of an entire society and "nation."

Arab intellectuals and reformers focused on the cumulative impact of individual endeavors of improvement. In an article in *Thamarat al-Funun*, the paper reprinted in full a speech given by one shaykh Muhammad Salih "al-Bahrayni" at a reformist school in Mecca, a lecture that repeatedly drew a connection between the time and life of one man and the collective fate of a nation or civilization. The school was well known in the Muslim world, its reputation extending far beyond the Arabian Peninsula. The "Madrassa Sawlatiyya" was the brainchild and lifelong project of Rahmatullah Kairanwi "al-Hindi," a South Asian Muslim who had fought in the 1857 rising in India and was subsequently forced to leave the subcontinent. After a protracted journey, Kairanwi ended up in Mecca, where he planned the opening of a school. To that date, the only education available in Mecca was the religious education offered at the Great Mosque. The school took its name from a benevolent Calcutta woman and was funded in part by contributions from Muslims on the subcontinent. It would become the first institution of higher learning offering both a religious and secular curriculum in Mecca.[55]

The Sawlatiyya school may have been an exceptional endeavor given its transnational character and the religious conservatism it faced at the holiest site of Islam. But it was by far not the only school that opened as part of reformist and revivalist efforts of Muslims and Christians to reinvigorate the Islamic and Arab civilization. Improving education was perhaps the most central effort by *nahḍawi* reformers. Several new schools were inaugurated in the Levant, Egypt, and Istanbul in the second half of the nineteenth century, some founded by missionaries, others originating in Ottoman or private efforts. Some of the publicists and journalists who established Beirut's most important newspapers directed and taught at a number of these new institutions of learning.[56]

It is fitting, then, that a reformist school in Mecca would have been the site for a lecture on "man's time, the benefits of saving it and taking interest in it, from the vice of wasting it." What is remarkable about the lecture is the extent to which the speaker emphasizes the malleability of a man's time and, therefore, one's life. The piece began by equating time with "a man's life, or a man's happiness."[57] It went on to detail how in his time (or in his life), man was able to be creative and industrious and to produce good results; "you plant and harvest, you work, invent, think, judge, ponder, and scrutinize, and your happiness lies in saving your time." The lecture continued, "What is the meaning of your life *(ḥayāt)* if you wasted your lifetime *('umr)* and lost your time *(waqt)*," and if the present circumstances lead you to calamitous future consequences? At this point you look at yourself and you find yourself to be nothing, as if you did not live one minute of your time . . . , which you lost." Again, the shaykh reminded his audience that time meant actively shaping one's life. "Time is according to what you wish for, if you want it is happiness for you and if you want it is bad luck," the speech claimed.[58] Time was an expression of the destiny lived, "and you live in it [time] when you seize its opportunity." If man in his time did memorable deeds for the greater good, "then your striving will be rewarded, and your trade will be a trade that will not be unprofitable." Wasting one's time, the shaykh warned, meant to give one's "psychological desires" free range, at the risk of walking past one's time, that is, perhaps failing to live up to one's true potential in life. With the strong emphasis on individual life and the ability to steer fate, the profit motive and talk of economic advantage serve as a metaphor more than an unambiguous statement in favor of capitalism and material gain.[59]

The speaker then moved from imploring individuals to save time to an assessment of contemporary Arab and Islamic civilization. He described "us" to be a people seeking knowledge, looking at the world with a curious eye and dedication. "Our time during the years of studentship is full of options, basic blessings, and ample utilities if we spend it on obtaining [knowledge of] the useful sciences" and "useful knowledge that enlightens our thoughts and polishes our brains, widens our perception, and guides us on an even path into a clear direction." Like this, the shaykh suggested, "we will be able to improve our [current] state and to reform our situation, to refine our morals and the education of our sons." Amid all his warnings and appeals, Muhammad Salih sent a clear message: individual energy spent on useful things and self-improvement was equally important in changing the fate of an entire people or civilization.[60]

Arab writers focused on fate and the ability to shape one's destiny when commenting on time. With *waqt* (most commonly used to denote clock time) and *zamān* (a more philosophical word for time with allusions to eternity, infinity), the Arabic language already uses more than one word for naming time. The strong interest among intellectuals and essayists in individual conduct and national destiny added another term to the debate, a notion perhaps best understood to mean fate in the sense of "the occurrences of a life's time span." *Dahr,* in Arabic, resonated with a religio-philosophical tradition that harked back as far as pre-Islamic poetry.[61] In the context of Arab and Islamic engagements with time, *dahr* became a vehicle for arguing that religion and modernity were reconcilable. Self-declared reformers were eager to demonstrate how compatible Islam was with modern scientific knowledge. *Dahr* was a concept that rendered individual efforts at self-improvement congruent with a certain interpretation of religion. Since individual efforts added up to a national accomplishment, the entire Arab and Islamic "nation" or civilization was able to transform itself without losing its faith.

In pre-Islamic Arab poetry, whatever afflicted man stemmed from *dahr,* the provenance of good and bad alike. Later in the Islamic tradition, time as fate became one of several powers that fell under God's absolute command. The near deification of time in pre-Islamic thought was now denunciated. God's unrestricted sovereignty over time and fate was eventually challenged by modernist thinkers like Jamal al-Din al-Afghani and Muhammad 'Abduh, protagonists of the *Nahḍa* in Egypt, who taught that Islam meant activity and that individuals were responsible for their actions as well as the welfare of the community. They summed up their program in a frequently quoted Qur'anic verse, "God changes not what is in a people, until they change what is in themselves" (13:10). According to al-Afghani, to believe in predestination was to believe God will be with a person who acts rightly.[62]

Notions of time as individual and collective fate found their way into yet another genre of Arab literary production in the late nineteenth and early twentieth century. Poems, and especially clock poems, conveyed a carpe diem motive that urged readers to make good use of their time on earth since life was short and youth a fleeting good. In one exceptionally long such poem titled "Youth and Time," published in three installments in the journal *al-Muqtataf,* the author, As'ad Daghir, implored time to slow down: "Lighten the march, O mounts of time; since my youth, I have not yet achieved my desires;" and "My eye does not want to see you, O Time [*zamān*], running away with my youth like a race horse." Youth was "the most pleasant of all phases through which man passes." "If life were spring," the author continued, then youth is more splendid than the month of April. Time, on the other hand, is portrayed as "an old, withered man, ancient in years; death and eternity are yours alike." And time threatened to clad everything in old age: "No particular place you visited, land or sea, that you did not penetrate over the course of seconds."[63]

Other passages of the poem admonished the reader to "leave aside your fascination with time [here more likely to be understood as fate], for what you hope from it is useless; drive yourself," and "accept advice that is sincere and pay attention, wake up, regain your consciousness, aware of what is with you,"[64] thus striking a similar tone as other essays and articles pleading contemporaries to actively steer their fate instead of being pushed around. And there was no "making excuses for yourself with 'I wish' 'maybe' 'perhaps', for these words are at the brink of the cliff of failure," for "Hope is beautiful, but striving . . . is its prerequisite, not laziness."[65] Another line asked, "Do you live in the shadow of futility hoping that youth will remain overshadowing you? So then make use of its minutes."[66] Spending time wisely above all meant spending it on useful activities. "He who in the time of youth passes his bright days asleep and wakes up for the pleasures of the night; Misery will inevitably rush to his abode: his regrets will compound and his woes will be visible."[67] Lines such as "spend your youth on that which brings benefit and utility, before your chance passes you by in vain," and "O you our youth, this is the time of your struggle [*jihād*]" struck a similar chord as other texts that encouraged Arabs to spend their time on "useful" activities.[68]

The connection between individual time and the fate of a nation was made most explicit in the following lines: "And strive for that which will save your country from the ignominy . . . it suffers from; From the ignominy of backwardness and crushing poverty; from the ignominy of wretchedness that has encompassed it from end to end; From the ignominy of ignorance that rules."[69] The "totality of calamities and maladies which encroached on our country are out of control," the author warned. Instead of indulging in self-deception, delusion, and confused dreams it was necessary to redeem the time left and "not allow a second to pass us by in vain."[70]

In the late nineteenth- and early twentieth-century world of the Eastern Mediterranean, the pre-Islamic poetical stance of carpe diem was revived to advocate wise time management and to caution against wasting time. Another example of blending a centuries-old form of art with contemporary, nineteenth- and early twentieth-century concerns was neoclassical Arabic poetry. The neoclassicists were a loose group of primarily Egyptian and Iraqi modernist poets who, at the turn of the twentieth century and beyond, wrote a particular genre of poetry that combined adherence to classical meters and forms with utterly new contents such as trains, telegraphy, and clocks.

One such clock poem was published in *al-Muqtabas* in 1908.[71] The author was Maʿruf al-Rusafi, one of Iraq's most famous poets of the first half of the twentieth century. Born in Baghdad in 1875, al-Rusafi became a teacher of Arabic language and literature before embracing a more actively political role in his writings. In the wake of the Young Turk Revolution of 1908, he celebrated the adoption of an Ottoman constitution and wrote more openly political journalism and poetry. Al-Rusafi lived in various exiles over the course of his life, moving between Baghdad, Istanbul, Beirut, and Jerusalem. He is best known for his political poetry directed against the British Mandate rule over Iraq in the interwar years. Al-Rusafi died in 1945 in Baghdad.[72] His poem "The Watch" (grammatically, a feminine word that can also mean "hour") depicts the mechanics of a watch whose ticking is likened to the movements of time and fate:

> Mute, her tongue can voice no word,
> except the deepening pulse within her body
>
> She speaks the dialect of "Tim-Taam," a meaningless cant,
> and is only eloquent on matters of time,
>
> An inner pulse by which she calls her lovesick heart
> in kind her heart responds to her,
>
> Through this beat run time's vicissitudes,
> by her blindness, the destiny of mortals becomes evident
>
> Signs and markers are scored on her face, by which
> the people are led to their appointed lots
>
> She walks within each moment, measuring it;
> for what is time if not her pace and step?

By her, the people exact their promises;
her guidance leads those who have strayed from time

She eats like anyone, like my brother in faith,
but what is her food but the twists of her entrails?

A clock hand revolves around her face like one off course
his landmarks lost within the darkness

Her turning reveals the sun's location
even when clouds have hidden the sunlight

More wonderful still, she is the result
of the thoughts and intellect of common folk

It was an easy thing for reason to build her;
her very construction was a matter of time

She summons the youth of the day with her strikes:
"Strive with the zeal of those who overcame my limits,"

"And don't neglect the times, for they cut
and tear apart the bonds of life with their blades."[73]

• • • •

THE RELATIONSHIP between individual exertions to improve one's fate and collective gain was perhaps expressed most clearly in a piece that appeared in Butrus al-Bustani's journal *al-Jinan.* The essay's author, Salim Diyab, started his piece by describing how ancient societies held different views on the prediction of fate and how to define certain times of luck or misfortune, without distinguishing between different aspects of time such as fate and life. The ancients failed to see the movements of time and fate as the ultimate source of self-engineered social mobility but rather, it was implied, believed in the almighty power of fate against which the individual stood powerless.[74] Diyab rejected such ideas as superstitious rumors. But he did acknowledge that a concern for the future, a proactive way of "encountering time," as Diyab put it, was mandatory for everybody who strove to succeed in life. Such an activist encounter with time emanated from "good planning and training" and devoting oneself to "what will happen before it happens."[75] If man made an intellectual effort, it was possible for him to limit his thoughts to earnestness, perseverance, hope, economizing and dividing up time, and studying local and international events, Diyab stated. Whoever observed such caveats and spent his time in useful ways was able to stop in front of his fate with a sturdy soul, not fearing its feckless turmoil.[76]

In a move characteristic of *nahḍawi* authors, Diyab then turned to history to corroborate his arguments. History and historical time were a popular and frequent subject of articles and other publications, interpreted to serve the purpose of the *Nahḍa.* Authors like Jurji Zaydan in his journal *al-Hilal* as well as several anonymous or less well-known writers talked extensively about different golden ages in Islamic history and the evolution of civilizations. Articles about history, historical ages, and how to launch a renewed golden age for the Arab and Islamic civilization were another expression of a general interest in time: in this case, historical time. The heightened interest in the nature and use of time and the development of a genre of history writing through an inquiry into the histories of the region were two sides of the same coin. History and the development of civilizations were linked to the progression of linear time in very literal ways. Arab intellectuals wielded the terms "progress" *(taqaddum)* and "backwardness" *(ta'akhkhur)* when comparing historical developments of "successful" and "laggard" civilizations. All Arab periodicals of the period carried countless articles expanding on these notions. When writers described the act of setting a clock or watch back or forward (as "Arabic time" necessitated on a daily basis), or adjusting a clock that was running slow or fast, the words for "back" and "forward," "slow" and "fast," were variations of the Arabic words for progress and backwardness.[77]

According to Salim Diyab, author of the piece in *al-Jinan,* the history of "civilized" nations provided a lesson for "encountering time," as the headline of his essay ran. When looking at the "civilized world" and the reasons for progression and backwardness in that world, Diyab held, it was clear that the progress of advanced nations was based on an improvement of their social situation, "which forced them to be in solidarity with each other." The author then turned to history, to the "days of the Caliphate," likely the time of the Abbasid caliphate in Baghdad from the mid-eighth to the mid-thirteenth century CE, commonly viewed as the Islamic Golden Age. In that time and age, according to Diyab, "the Arab nation [al-'umma al-'arabiyya]" advanced "when it sought to improve its works."[78]

When the Mongol invasion and the sack of Baghdad ended this golden age of productivity and pride, international events "played with them [contemporary witnesses to these events]," an example of being pushed around by fate rather than shaping one's destiny. The Arab nation went astray and declined after it had been independent. Arabs learned to live with humiliation and got used to it. "Our Arab nation," Diyab deplored, "since it fell behind in our days, has only been famous for idleness and laziness, despite the fact that some others who fell were on guard and got up again and regained what they lost." Diyab's emphasis on the evolution of civilizations as passing through different stages explains why a combination of Darwinist and Spencerian evolutionary sociology held a powerful grip on

Arab audiences at the time. If social progress followed certain laws, those could be studied and steered at the level of the individual and society as a whole. As a social process, evolution was thus a wellspring of strength and a tool for self-improvement if these laws were followed.[79]

Diyab's essay contained a phrase pivotal for understanding the *Nahḍa*'s relationship to the "time is money" motive. In a strong and successful society, Diyab held, everybody was working "for the prosperity of the nation whose prosperity is dependent on the fortune of her individuals." Whether it was the productivity of the English worker who enjoyed a healthy breakfast or the industriousness of a society that knew about time management and dividing up time according to tasks, individual efforts added up to a nation of studious people. Such a nation as a whole might be able to withstand even the challenges posed by European and American imperial aspirations in the Eastern Mediterranean. The notion that individual efforts benefited the nation was a widely used trope in *Nahḍa* texts. When the Syrian Protestant College awarded a prize to a local watchmaker, the phrase even found its way into the awardee's diploma.

Elias Ajiya, the watchmaker to be feted, was a local celebrity. Hailing from an Aleppine Syriac-Catholic family, he seems to have resided in Beirut in the latter decades of the nineteenth century. Ajiya was known to be an inventor of all kinds of curious apparatuses, but watches were his specialty. Ajiya's works, whether a fan or a device for heating water, were the topic of several newspaper articles.[80] His "planetary clock," really a clock-cum-calendar, garnered him an invitation to Paris, where he presented his arts and crafts piece in front of the Paris Geographical Society. The clock was composed of a chandelier carrying little globes as representations of the earth, the moon, and the sun, all of which rotated in accordance with their actual movements in the sky (the earth moved around the sun, the moon orbits the earth during one month, and so on).[81] *Thamarat al-Funun*, the newspaper writing about the prize ceremony at the Syrian Protestant College, reprinted Ajiya's degree as stating, "When it was mandatory for the entire nation to have the sons of the nation make an effort and strive for everything that pertains to the progress of the nation, Elias Ajiya was among those who spent their time on useful inventions for the nation."[82]

The exhortation to heed the calling, pull oneself up by the bootstraps, make good use of one's valuable time, and through such self-improvement contribute to the health of the body politic overall, was a leitmotiv of the *Nahḍa*'s plea for the self-induced invigoration of Easterners. In a largely overlooked instance of translation and appropriation, this particular element of the *Nahḍa*'s self-improvement program had been borrowed from the Victorian champion of all self-made men, Samuel Smiles, who in 1859 had published a global best seller by the title of *Self-Help*. His book (which would establish the genre of self-help books) contained snapshots from the lives of successful entrepreneurs, businessmen, and engineers who had raised themselves from a modest socioeconomic background to material success. His Arab readers, however, focused not on material wealth but on the connection between individual and national success, a causal relationship established many times before in reformist writings.[83]

Viewed from this perspective, Smiles's chapters and sections about "perseverance," "energy and courage," "knowledge as a means of rising," and "punctuality" offered his Arab readers a tableau of remedies against almost everything that had vexed them. Most noteworthy, his chapter on "business men" talked about the value of time. "Men of business are accustomed to quote the maxim that time is money, but it is much more; the proper improvement of it is self-culture, self-improvement, and growth of character. An hour wasted daily on trifles or in indolence, would, if devoted to self-improvement, make an ignorant man wise in a few years," Smiles stated. Such conduct enabled individuals furthermore to push rather than to be pushed around, meaning "to get through business and carry it forward, instead of being driven by it." Smiles's manual of instructions also came with a warning: "Some take no thought of the value of money until they have come to an end of it, and many do the same with their times. The hours are allowed to flow by unemployed, and then, when life is fast waning, they bethink themselves of the duty of making a wiser use of it." Since time-profligate individuals were so wedded to bad habits, they were bound to fail. "Lost wealth may be replaced by industry, lost knowledge by study, lost health by temperance or medicine, but lost time is gone forever," Smiles explained.[84]

Smiles laid out the connection between individual and collective success clearly on the first page of his oeuvre: "For the nation is only an aggregate of individual conditions, and civilization itself is but a question of the personal improvement of the men, women, and children of whom society is composed. National progress is the sum of individual industry, energy, and uprightness, as national decay is of individual idleness, selfishness, and vice."[85] Here, Arab readers found what they had been looking for, a plausible way of thinking about national progress as the sum of individual efforts at time management and time saving and other improvement endeavors. Ja'qub Sarruf, one of the editors of the popular journal *al-Muqtataf*, requested permission from Smiles's Scottish publisher, John Murray, to translate *Self-Help* as early as 1874. The Arabic version that finally came out under the title "The Secret of Success" in 1880 circulated widely in

Beirut and Cairo. The Egyptian ruler, the Khedive, had excerpts from *Self-Help* inscribed on the walls of his palace, next to verses from the Qur'an. Egypt even had a *Self-Help* society.[86]

Jurji Zaydan, a further *Nahḍa* figure and editor of the journal *al-Hilal,* wrote an (unfinished) autobiography that almost reads like a depiction of the sort of self-made man Smiles was praising: Zaydan described raising himself from a modest background of illiterate parents and gradually climbing the ladder through various jobs to end up as a student who excelled at the Syrian Protestant College in his studies of medicine. Upon discussing his interest in acquiring knowledge in the sciences, Zaydan states, "I had read parts of the book 'The Secret of Success' which Dr. Sarruf had translated into Arabic. Vigor and zeal sprang up in me; I read, as I said, some of it but was unable to finish the rest. Too great was the enthusiastic impact it had upon me to read about the lives of men who reached highest achievements by their own diligence and efforts and selfreliance [*sic*]. Amongst them, barbers and shoemakers, servants, artisans and maids who rose through their eagerness and viligance [*sic*] to the station of great people."[87]

Arabic was by no means the only language into which *Self-Help* was translated. According to Smiles's son, in 1912, translations existed in Armenian, Bengali, Chinese, Croatian, Czech, Danish, Dutch, French, German, Gujarati, Hungarian, Italian, Japanese, Marathi, Norwegian, Polish, Portuguese (made in Brazil), Russian, Siamese, Spanish (one from Buenos Aires as well as a European-Spanish edition), Swedish, Tamil, Ottoman Turkish, and Welsh. While the list is impressive enough, it is likely that many more foreign versions existed. This was an age in which notions of intellectual property and copyright protection were only slowly gaining a foothold, with enforcement being virtually impossible. Smiles's publisher, the Scotsman John Murray who had made a name for himself by acquiring the rights to several other nineteenth-century blockbusters (he published Darwin's *Origin of Species*), complained frequently about the several unlicensed translations of *Self-Help* (and other Murray publications) he knew to be circulating worldwide.[88]

Smiles's works were commercially even more successful in British India than in the Levant. As early as 1869, Smiles received a letter from a "Madras gentleman" requesting permission to reprint passages from his works; in 1912, translations were available in Gujarati, Hindi, Marathi, and Tamil, and likely Urdu as well. Some of Smiles's other works, "Duty" and "Character" in particular, were equally popular on the subcontinent. Schools were the primary audience for his works. The headmaster of a school in Dacca requested permission to compile a "best of" reader of Smiles's writings that would contain a few pages from each of his works. One Calcutta bookseller looked to compile an English-language reader of Smiles's work, 500 copies of which would be distributed to heads of schools in Bengal to be used in English-language instruction. Around 1906, this remarkable interest from the subcontinent prompted the publishing house *Murray & Sons* to bring out a special edition for the Indian market, produced from slightly cheaper paper and material.[89]

The global success and appeal of Smiles's message, whether in the Levant or British India, cannot be explained merely with the attractiveness of economic success and material gain. Within the mid-century Victorian context of its publication in Britain, Smiles's message may have indeed been intended as a statement in favor of laissez-faire economics and economic liberalism, as he is often interpreted.[90] But to his Arab readers, Smiles provided a recipe for national self-strengthening through individual time management and generally individual effort. The fact that Smiles was widely popular among Meiji reformers in Japan suggests a similar interpretation in that country. Since the beginning of the Meiji Restoration in 1868, Japan had, after all, embarked on its own program of strategic and selective adoption of Euro-American science, law, military organization, and more. *Self-Help* as national self-improvement was a global phenomenon. Smiles offered leading heads around the globe the prospect of modernizing from within. In places such as Beirut, time thus took on national and civilizational functions and meanings.[91]

Beirut and the Levant occupied a space in between, where creeping colonization was a real threat but had not yet taken on the concrete form of a British occupation as in Egypt. In late nineteenth- and early twentieth-century Egypt, a similar fascination with time and "the value of time" was widespread. In 1911, Ibrahim Ramzi, an Egyptian author, published a book that echoed the Arabic title of Smiles's *Self-Help* (*Sirr al-Najah,* which translates as *The Secret of Success*). The book, *Asrar al-Najah* (*The Secrets of Success*), featured a section titled "time is precious" *(al-waqt thamīn)* as well as advice on successful undertakings in agriculture, industry, and trade. Ramzi frequently pauses to move from general advice to commentary on what had allowed the "foreigners" in Egypt to come to control these sectors and what was required of Egyptians to force them to abandon their dominant position. But with the British seemingly there to stay, Egyptians more frequently and openly criticized technologies like railways and trams and the industrialization of time they stood for.[92] In the Levant, commentators were less critical and instead urged fellow citizens to turn European "weapons" more openly against imperial encroachment. Even between Beirut and Egypt, concrete local political conditions and the differences between them defined the terms of the globalization of time.[93]

• • • •

In some instances, what Arab authors desired for their own society was a strategic emulation of Euro-American temporal practices; in others, trajectories of circulation were more complicated when subtle changes in terminology suddenly opened up a whole new register of thought that originated locally and resonated with Arab and Islamic societies in ways no European concept ever would. Curiosity about time display, time management, and truthful indications of time, as expressed in newspaper exchanges over sunset times, belie the image that Arab authors painted of their own society. Rather than temporally oblivious, at least the reading and writing public took time seriously. Arab reformers were as concerned with the waste of time as William Willett was with the waste of daylight. The theme of the time-profligate Easterner sprang more likely from the imagination of Euro-Americans themselves. The plurality of time that characterized global cities such as Beirut called for comparisons, for examinations of different ways of measuring and using time. A whole array of print products—newspapers, magazines, and countless little brochures and pamphlets—once more functioned as transmission belts for ideas and information about the times of others. Local intellectuals, journalists, editors—reformers tout court—latched onto every bit of information about "British" or French innovations in timekeeping and temporal habits they were able to secure. Based on such scouring, reformers subsequently tailored these contents to fit the conditions on the ground in the Levant and elsewhere. These conditions strongly suggested the need for catching up, for positioning the "Eastern" civilization closer in time to the space occupied by Europeans, not least with the help of time management. Time—hybridized and adapted—attained meaning only through comparisons to other, global times.

CHAPTER SIX

# Islamic Calendar Times

In the fall of 1910, religious learned men from Damascus to Algiers were following contemporary media even more attentively than usual. The world of muftis, judges, and scholars had been jolted by a series of missteps in determining the precise end of the fasting month of Ramadan; the ensuing debates foregrounded a complicated balancing act of attempting to reconcile technology, science, and religion. These discussions were part of a broader struggle over the interpretation of Islam, pitting reformers against conservatives on a range of matters across the Muslim world; when such debates concerned matters of religious worship, emotions ran high. In the late nineteenth and early twentieth centuries, these quarrels widened a growing divide between proponents and opponents of adopting Euro-American lifestyles and inventions. Reformers and intellectuals in the Levant may have advocated the targeted adoption of certain elements of Euro-American time, material and other. Among some more religious-minded thinkers, such and other emulations and translations did not go uncontested. During those very same years when Arab writers exhorted their contemporaries to heed the value of time, another aspect of time, the Islamic calendar, became the focus of intense debates in the Eastern Mediterranean and other parts of the Muslim world. While the 1910 Ramadan clash was by no means the first such incident, the scope of interventions in this instance indicated a remarkable degree of polarization over a topic that previously might have intrigued only a couple of bookish types.[1] What had happened? In this particular case, what lay at the heart of the confrontation was a controversy over adjudicating religious calendar time.[2]

The dispute centered on the question of whether it was in accordance with Islamic law to use the telegraph in reporting the news of sighting the new moon, and whether it was permissible to rule on the beginning or end of Ramadan based on such a telegraphic message. What Muslim scholars found most difficult to accept in incorporating the telegraph into the legal framework of adjudicating calendar time was the questionable accuracy and reliability of apparatuses and machines and, therefore, trust in modern technology. Whether and why the telegraph could be trusted to transmit proven truths rather than errors and lies would become a focal point of disagreement. What is more, discussions about telegraphy and the timing of Ramadan inevitably raised questions similar to those preoccupying Euro-American reformers of time in the same years. The broader, more portentous issue looming over the telegraph controversy was about the uniformity of religious timing practices. Was it possible for Muslims to agree on the same calendar, and would the telegraph aid such uniformity? Most of all, in an age when telegraphs, steamers, and railways seemingly made distances disappear, what did "Islamic time" even mean? Was it time for all Muslims in the entire world or those living within a single country or region? The controversy over the Islamic calendar thus raised the problem of just how universal time could be. Potentially, as some would come to argue, the uniformity of timing practices offered contemporary Muslims a semblance of unity. In an age when especially religious thinkers had come to conclude that their lifestyle was soon to become extinct by modernity, matters of unity, consensus, and solidarity acquired heightened importance.

From the Eastern Mediterranean to North Africa, some of the most prominent scholars of their generation issued fatwas or wrote legal tracts weighing in on telegraphy, many repeatedly, some in the form of entire books devoted to the topic. Among them were well-known figures in the history of the Middle East and Islam such as the Damascene Jamal al-Din al-Qasimi and Rashid Rida in Cairo. Their arguments as well as those advanced by less prominent figures constitute a prime example of how Islam gradually incorporated new technologies such as the telegraph and modern clocks and watches. They illustrate Islam's ability to render itself compatible with change and to adapt to evolving circumstances.

• • • •

After a long, hot summer in 1910, Ramadan, the holy month of fasting for Muslims throughout the lands of Islam, was about to end. According to the Islamic tradition, God revealed the Qur'an to his Prophet, Muhammad, during that month, and Muhammad had ordered his followers to fast from dawn to sunset. ʿĪd al-Fitr, the holiday of breaking the fast, celebrates the end of Ramadan and the beginning of the following month. Ramadan 1910 had been like any year's holiday: life slowed down as people tired from their lack of sustenance during the day; businesses observed reduced hours; Muslims enjoyed the company of families and friends, gathering to celebrate the fast-breaking meal together at the end of the day.

On the twenty-ninth of the month of Ramadan, the chief judge of Egypt assembled "at the shariʿa court with the learned men of Egypt," among them the head mufti of Egypt and the leading scholar at Cairo's renowned al-Azhar University. This was the day and following night when the new moon was likely to be sighted and the fast-breaking holiday would begin. The group waited until the evening hours—but no witness came forward to report seeing the new moon. The following day, the governor of Cairo—a representative of Egypt's political leader (the Khedive or viceroy)—received a telegram from the administrative head of the southern Egyptian province of Aswan: the shariʿa court there had confirmed a sighting of the new moon. A witness in Aswan had testified to seeing the moon, and a local judge ruled that ʿĪd al-Fitr would start that day.[3] In Alexandria, Muhammad Bakhit "al-Mutiʿi," then a judge at the shariʿa court in the Mediterranean city and an important scholar later in life, also declared the holiday based on the Aswan sighting.[4] He went so far as to order the ritual shooting of cannons and lighting of the candles that were affixed to minarets, thus officially announcing the beginning of the ʿĪd.[5]

But doubts arose immediately, even among the administrators and scholars themselves.[6] The religious authorities still gathered at the chief shariʿa court in Cairo continued to wait without ruling or acting; the Khedival governor in Cairo forwarded the telegram announcing the Aswan sighting to the qadi of Egypt. Traditionally, the head judge of Egypt sent telegrams to judges throughout the country announcing the fast's beginning and end.[7] Now, however, the qadi of Egypt declared he lacked the authority to rule under such circumstances.[8] The governor of Cairo—a political figure rather than a religious one—eventually went ahead and ordered the cannons fired for all of Egypt, officially signaling the holiday without awaiting a ruling by the chief judge.[9] That decision, too, was challenged as soon as it was announced. "To many esteemed observers transmitting the ruling of the judge by telegraph and acting upon it appeared doubtful," as Bakhit himself later observed.[10] The telegraphic back-and-forth, highlighting the spat between state officials and religious scholars, prompted many Muslim legal thinkers to ask fundamental questions about the use of technology in religious rituals. Was it in keeping with the stipulations of Islamic law to simply send a telegraphic message about somebody's crescent-sighting and officially act upon it—that is, to rule on the beginning of Ramadan or the fast-breaking holiday?

Textual evidence that could shed light on such questions was limited. The most widely agreed-upon relevant source was a few lines from a number of texts that most experts deemed authentic. According to these reports, the Prophet Muhammad had said something along the lines of, "fast upon seeing it and break the fast upon seeing it [again], and if it is cloudy, complete the number of days as thirty." "It" was the crescent of the new moon. All this meant was that Muslims were to start fasting when the new moon of Ramadan was sighted and to cease fasting when the crescent of the following month, Shawwal, became visible. If weather conditions complicated the act of sighting, it was mandatory to break the fast once thirty days had been completed, counted from the moment when the new moon of Ramadan was first spotted. But scholars could not entirely agree on the precise meaning of Muhammad's instructions; what is more, as in so many other instances, some contested the textual evidence for what it omitted, and for what had simply not been part of Muslim daily life in the seventh century CE.

At the turn of the twentieth century, then, religious scholars found themselves embroiled in discussions of how to treat a telegram reporting crescent-sightings. What was at stake in these debates was not just the question of using technology in religious matters but also the relationship of political to religious and judicial authority. In the Ottoman lands as well as in Egypt at the turn of the twentieth century, the state appointed the chief judge, or qadi.[11] Besides the qadi, there was the mufti, literally the "issuer of fatwas," best understood perhaps as a consulting jurist. When the qadi needed to issue a ruling on a particularly difficult or unprecedented matter, he turned to the mufti—schooled in legal theory—for advice in the form of a legal opinion or fatwa.[12]

In determining the religious calendar times of Ramadan, a range of legal personnel interacted with the state's bureaucratic establishment. When a witness spotted the new moon, he was obliged to testify in front of a judge, who would then rule that the month of fasting had begun or ended. Muftis might have previously issued fatwas on Ramadan or the fast-breaking holiday, which the qadi could consult in his decision. Eventually, the political establishment would declare Ramadan to have begun or ended, and local or regional governors would announce the news by ordering cannons fired and candles lit. These exchanges between officials of the state and religious personnel could be fraught at times. Britain had occupied Egypt beginning in 1882 but left the political structure of the Khedival regime partially in place. Allowing telegraphy to be deployed in the timing of Ramadan could be construed as an instance of Europeanization. Under rapidly changing political circumstances, people naturally viewed any conflict over the religious calendar as a struggle over authority.[13]

The controversy over the timing of Ramadan was also a complicated epistemological conflict about the foundations of Islamic law and the methods of establishing and deriving laws, historically as well as in the present. At core, using the telegraph to communicate the news of the moon sighting was a question of trust and certainty. To Muslim scholars versed in traditional legal theory ('Uṣūl al-Fiqh), talking about trust, certainty, and reliability of reports and messages resonated with a problem long discussed in a different context. As long as God's Prophet, Muhammad, walked the earth, his example, the sunna, guided Muslims in their beliefs and practices. When after Muhammad's death in the seventh century storytellers began to informally circulate reports about sayings and actions attributed to the Prophet *(hadith)*, it was suddenly necessary to establish sound criteria in investigating and evaluating the origins of such reports. By the eighth and ninth centuries, an increasingly sophisticated "science" of hadith proofing had established itself.[14]

The Arabic word for such messages about Muhammad's words and deeds is *khabar*. Today, *khabar* is most commonly wielded to mean message, often in connection with news journalism. When Arabic television channels announce the equivalent of what would be "breaking news" on a US channel, they often display a red banner reading *khabar 'ājil*—urgent message. To late nineteenth- and early twentieth-century Muslim jurists and muftis, the analogy to draw here was a different one: the legal status of the telegram or telegraphic message *(khabar)* was analogous to reports about Muhammad's sayings and actions. Whether it was permissible to use the telegraph in transmitting the news of the moon sighting and to issue a ruling based on such a message therefore required classification and evaluation as prophetic reports that pertained to the Islamic scriptural tradition. According to those scholars who approved telegrams and deemed them trustworthy and reliable, a telegraphic report held the status of what legal theorists termed a "solitary message," a text that had been handed down by just one channel of transmission instead of recurrent reporting by numerous people.[15] This meant that such a solitary message was of somewhat diminished certainty and value, yielding probable rather than absolute knowledge.[16] Some disagreements remained unresolved but all in all, to those advertising the reliability and trustworthiness of the telegraph for determining the Islamic calendar, the telegram legally was analogous to an authentic report about Muhammad's life. Debates about the role of new technologies such as telegraphy in religious timing rituals were clad in established legal terminologies to be eventually integrated into Islam.

• • • •

In the fall of 1910, newspapers from the Levant and Egypt had reported on what many regarded to be botched rulings and nonrulings on

Ramadan 1910. One such article caught the attention of Jamal al-Din al-Qasimi, a prominent religious scholar and reformer in Damascus. His investigation into the matter and later writings on the subject triggered what he referred to as "this legal renaissance," a slew of fatwas, tracts, and newspaper articles centering on the timing of Ramadan via the telegraph. Al-Qasimi (1866–1914) was a learned man or so-called *ʿālim* (pl. *ʿulamāʾ*) who was one of the leaders of the local salafiyya reform movement.[17] Islamic reform in late Ottoman Syria and Egypt advocated a return to the practices of the "pious ancestors" called *al-salaf al-ṣaliḥ* in Arabic—hence, salafiyya. Indeed, these reformers adopted a scripturalism that stressed the authority of the textual sources of the Qurʾan and the sunna for determining orthodox beliefs and practices.[18]

The salafiyya movement was propelled by two broader concerns. European encroachment and imperialist pressure on Muslim rulers seemed to suggest that the customary interpretation of the sources of the Islamic faith no longer served as a reservoir for Muslim inspiration and vitality. Centuries of legal and interpretative practice were now deemed distorting imitations *(taqlīd)*, under which the true and original tenets of the religion lay buried. Moreover, the threat of European conquest and colonization suggested that Muslims needed to overcome their differences and unite in the face of European imperial threats. The simplicity of an original and undiluted religion would minimize sources of disunity.[19] In the views of nineteenth-century Muslim salafi reformers, a simpler Islam rendered the religion applicable anywhere, at any moment in time. Using modern telegraphy within a classic shariʿa framework, as al-Qasimi and others advocated, supported this notion of a universal religion and became an argument for Islam's compatibility with the circumstances of all times and ages. When Jamal al-Din al-Qasimi wrote about the telegraph in determining Ramadan, he argued that the use of technological innovations like telegraphy within a shariʿa framework "proved that Islam is a religion that supports civilization and marches alongside it." In this vein, al-Qasimi promised to demonstrate that the "faculties and rules [of the shariʿa] had descended generally to be applied to the necessities of all times."[20] In al-Qasimi and other Muslim intellectuals' views, it was possible to be both Muslim and "modern."

• • • •

In 1910, Jamal al-Din al-Qasimi penned an article titled "Good News about Science" in which he explained that he had previously conducted research on the issue of the telegraph and, spurred by the recent controversy in Egypt, had decided to share his findings. His inquiries had brought to light that in the past, other well-known scholars had issued approving fatwas about using the telegraph in the context of religious rituals, hence the "good news." Conservative opposition to Jamal al-Din al-Qasimi and his fellow reformers' ideas was strong. Al-Qasimi and those opposing him competed to outdo each other in laying bare the insufficient and erroneous nature of the telegraph or alternatively in praising its official, government-vetted character. Different Islamic legal schools had established strict criteria for the transmitters of valid Prophetic reports; among all of them, being trustworthy, reliable, and of "moral rectitude" [*ʿadl*] stood front and center. Now it had to be decided whether the telegraph could be said to fulfill these requirements.[21]

Conservatives in Damascus established their own newspaper: *al-Haqaʾiq* took on salafis and Westernizers at the same time, often painting the salafis as just another group of Europeanizers who preached reform when all they really meant was the adoption of Western science and lifestyles.[22] To the conservatives at *al-Haqaʾiq*, relying on the telegraph was a "corrupted" practice. "How is it permissible for us to act upon the telegraphic message when it is a line of doubt and a path of ambiguity?" the paper asked, using the Arabic word "āya" for line, deployed to describe a line in the Qurʾan, which was literally understood to be a "sign" of God. The word "sabīl," used for path, also had a religious ring to it. In *al-Haqaʾiq*'s critique, the religious meaning of "āya" and "sabīl" was literally corrupted by the telegraph's frequent misspellings. The conservative paper asked, "Is it permissible for us to build the rulings of our religion on the telegraph given how much people are thrown in confusion due to the telegraph? Given how often they understand the opposite of the intended meaning?"[23]

To *al-Haqaʾiq*, a telegram could not be relied upon, and "the claim that it was a proof [bayyina]" was incorrect since the telegram "is in contrast with the truth."[24] The rulings of Islam and the sharīʿa could not be founded on "void analogies," referring to the parallels drawn between the legal category of a "solitary message" about the Prophet's life and a telegram. Moreover, to accept a message from a "deaf apparatus" such as the telegraph was to abandon what the Prophet had done, and that was, to hear a human being, a witness coming forward to testify about sighting the moon in his own voice.[25] To *al-Haqaʾiq*, the governor's unilateral action in shooting the cannons without awaiting a ruling by a religious court illustrated the tensions between religious and governmental claims to authority. The conservative authors at *al-Haqaʾiq* saw this as further support for their suspicion that Europeanized political authorities denigrated the role of religion and had all but sold out their convictions to the West. If the announcement of the impending end of the fast occurred without a sharīʿa ruling, *al-Haqaʾiq* warned, if instead it was "merely the opinion of the governorate and its approval, then we fear that . . . the rulings of the religion will be in the hand of the ignorant [*jāhilīn*] [political] rulers."[26]

Stoked by the repeated retorts from his conservative opponents, Jamal al-Din al-Qasimi began to inquire more thoroughly into the matter. He must have been astonished: under every stone he found yet another scholar, judge, and mufti who had ruled on using the telegraph in adjudicating the religious times of Islam. Even while he was still trading blows with the authors over at *al-Haqaʾiq,* al-Qasimi announced his intention to compile all these materials into a book.[27] The result was a roughly hundred-page tract titled *Guiding Mankind on Acting Upon the Telegraphic Message,* published in 1911. It detailed what different legal traditions had to say about so-called solitary and recurring Prophetic messages, trustworthy transmitters, proper evidence in Islamic law, and the like. A historical section covered older methods of sending messages, such as using courier pigeons or camels; at the book's end, the Damascene scholar and reformer appended materials written by others about telegraphy. The appendix began with two poems about the telegraph, written in the classical form of Arabic poetry, the so-called *qaṣīda,* and followed by a total of twelve fatwas from contemporaries around the Eastern Mediterranean who had opined on using the telegraph in connection with determining the calendar times of Ramadan.[28]

Jamal al-Din al-Qasimi made it clear that this was not the first time Islam had incorporated recent innovations. Islam was a flexible way of life, set up to adapt to changing circumstances: "It belongs to the wisdom of the system of law that it did not make explicit all [its] practical applications because matters changed," al-Qasimi explained.[29] "The introduction of the telegraph resembles earlier innovations that did not exist in the time of the Companions or the Successors" of the Prophet, among them cannons and clocks for announcing the fast and timing the prayer.[30] Al-Qasimi even backed up his assertion with a verse from the Qurʾan: "There will be created what you do not know."[31] If anything stood in the way of using technology in determining religious time, Islam and Islamic law could not be held to be it.

• • • •

Al-Qasimi's book *Guiding Mankind* stressed the official nature of the government-operated telegraph and its long-standing service to rulers as proof of the technology's reliability. In part, this was a consequence of the proliferation of telegraphy. With the extension of Ottoman telegraph lines into the more remote parts of the empire and the presence of British and other line operators in Egypt, sending and receiving telegrams became a daily operation and experience for a growing number of Muslim bureaucrats and judges. The telegraph had made inroads into the Ottoman Empire ever since the Crimean War of 1853–1856, after which the network expanded rapidly to include some 320 offices around 1869. In the Levant, the first line was introduced in 1861 between Beirut and Damascus, with further connections soon linking these cities to Istanbul, Anatolia, Palestine, Baghdad, and the Persian Gulf. Egypt gradually developed into a telegraphic hub in the 1860s, when a stable connection linked Malta (and thus Europe) and Alexandria, and a Red Sea cable stretched from Suez to Karachi in British India.[32]

Rulers communicated only in serious matters concerning the well-being of the state, al-Qasimi argued, and since they had to sign with their names, errors such as misplaced diacritical marks or spelling mistakes and a distortion of the intended meaning were highly unlikely.[33] To corroborate his arguments, Jamal al-Din al-Qasimi laid out the Ottoman state's established legal requirements and standards for telegraph companies operating in its realms: telegrams had to be submitted in person to the telegraph bureau, signed and dated;[34] a telegram by a person unknown to the bureau and without proof of identity was inadmissible; when a telegraph was submitted by a substitute, that person's identity had to be verified as well; telegrams had to contain clear and intelligible content. A vetting process even existed for telegraphy clerks: two people needed to confirm that an applicant for a telegrapher position was of sound mind and had not been accused of any crime. He had to give a demonstration of his skills, too. And once appointed, the operator faced harsh punishment and degradation if caught committing forgery and the like.[35]

According to al-Qasimi, this "heavy responsibility" represented a sufficient guarantee against clerks making amendments to texts.[36] The entire Ottoman law regulating telegraphy entailed over seventy paragraphs of regulations, and such controls and scrutiny had made the telegraph a vital ally of kings and sultans. But more: merchants, even the pious ones, relied on the telegraph as well. They did not question telegraphic transactions—indeed, most of their purchases, sales, and donations were "built on telegrams from their partners and agents in [foreign] countries." The telegraph was "the pillar of trade and transactions today."[37] Better even, religious authorities were known to send telegrams as well. How often, al-Qasimi asked, did the highest religious official of the Ottoman state, the shaykh al-Islam, send a telegraphic message to the empire's provinces to announce the dismissal or appointment of judges to the shariʿa courts? It did not end at the level of local religious officials: the Ottoman sultan was simultaneously the highest Muslim religious leader on earth in his function as caliph. The caliph's ascension itself was announced to the remote regions of the realm via telegraph, and upon receiving such news people all across the Ottoman Empire celebrated. According to al-Qasimi, such widespread acceptance by government officials and ordinary people alike constituted a form of "consensus," which according to Islamic legal theory was a form of truth beyond dispute.[38] Officials and private persons sent telegrams to spread the news of someone's death; upon receiving such a message, they prayed for the deceased. Hence, in matters of religious ritual, the telegram was equally relied upon.[39]

Other experts followed Jamal al-Din al-Qasimi in recounting the long list of successful telegraph users. Muhammad Ibn ʿAqil al-Hadrami "al-Hussayni," a legal theorist based in Singapore, was following the 1910 events in Egypt apparently in the Arab-language press, just like al-Qasimi. He twice wrote to him, welcoming the latter's commentary on the question, since telegraphy and the timing of Ramadan were "matter[s] of frequent recurrence."[40] Ibn ʿAqil al-Hadrami related from Singapore that the Islamic Union, likely a local religio-political learned society, had taken up the topic of determining the beginning and end of Ramadan. He attended the meeting to state his opinion that a message sent via the official government telegraph should serve as a basis for action and for ruling on the month of Ramadan. The fact that the people of Egypt already relied on the telegraph in determining Ramadan should be an encouragement, he stated. Ibn ʿAqil's own views were clear: he contended that given the extent to which "people relied on the telegraph for civil transactions, announcing a death, severing diplomatic relations and declaring war, appointing and dismissing personnel," there could be no serious doubt that a telegram was trustworthy.[41]

• • • •

A GOVERNMENT-VETTED telegraph system may have convinced many but not all contemporaries of the technology's trustworthiness. For those still harboring doubts, Jamal al-Din al-Qasimi and his fellow reformers crafted another set of arguments that was built upon an abstract understanding of the telegraph as a medium. Categorizing the telegram as a "mere" medium allowed advocates of the telegraph to cast aside some of the shariʿa requirements for those reporting on the moon-sighting. In their diatribes against al-Qasimi and his ilk, the conservatives clamoring in *al-Haqa'iq* had founded their rejection of telegraphy on the notion that a telegram legally *was* the so-called shahāda, the testimony given about the sighting of the crescent, which was then sent to a judge who would move to rule on the beginning or end of Ramadan. The shahāda and the witness who gives testimony are held in high regard in Islamic law and serve as key instruments in establishing the truth in disputes of different sorts. Jamal al-Din al-Qasimi and others who favored telegraphy argued that the telegram itself had nothing to do with the testimony; it was merely a medium, not a testimony, and in a way not even a message. A telegram was a carrier of information and not a testimony, and the telegraph was not a witness.[42] The shariʿa's several rules for testifying and criteria for the witness himself were thus invalidated in the context of telegraphy.

Al-Qasimi stressed that Muslim rulers of past centuries had similarly relied on courier pigeons and postal camels in distributing urgent messages, and the telegraph was simply another such instrument of accelerated communication. Obviously nobody would want to argue that the camel was the testimony.[43] Indeed, al-Qasimi was keen to demonstrate that acting based on the telegraphic message did not even pertain to the broader legal context of testimony or shahāda—"the wire itself is not a witness of the sighting of the moon or the ruling of the judge," he clarified. It was a trusted message *about* someone who had given testimony to sighting the crescent.[44] As al-Qasimi pointed out, the Prophet Muhammad himself had received and accepted letters delivered and handed to him—even by non-Muslims who served as ambassadors and such.[45]

Muhammad Bakhit, the Alexandria judge who was involved with the 1910 controversy in Egypt, had a similar theory about the telegraph as a medium. Like Jamal al-Din al-Qasimi, he devoted an entire book to the telegraph, published around the same time.[46] When Muhammad Bakhit learned that al-Qasimi was completing a book on the same subject matter, he gifted his own publication to his Damascene contemporary and the two exchanged views.[47] Bakhit was born in the Egyptian city of Asyut in 1854 and later studied at Cairo's al-Azhar University. Bakhit recounted a number of events that had led him to author a book on timing Ramadan with the help of the telegraph. One was the controversy from Egypt in 1910; another was related to him by a friend of a friend from the Saudi Arabian city of Medina who was traveling in India. There, the friend encountered a scholar referred to as "shaykh ʿAbd al-Hayy," possibly ʿAbd al-Hayy al-Luknawi, a scholar of Islam from India.[48] The Indian shaykh handed the traveling friend a letter for Muhammad Bakhit, approaching the well-known scholar with the following account: across the subcontinent, several instances of "disturbance and differences in trusting the telegraphic message about the fast and ending the fast" had been registered. The question was whether a telegram announcing that the crescent had been verified in one district was a basis for action in another, and what to make of the fact that in India, the telegraph clerks were mostly non-Muslim.

Like Jamal al-Din al-Qasimi, Bakhit conceptualized the telegraph as a medium that had to be understood in the legal and physical context of other media.[49] The phonograph, for instance, which was slowly making its entry into the world of the Eastern Mediterranean in this period, repeated the words of a speaker in the sound of his voice, hence the origin of the message was beyond doubt. The telegraph was different, but it too contained means for safeguarding against falsely claimed identities and forgery, most importantly the sender's signature and name on the message originally delivered to the telegraph office. The mailman and the telegraph operator were both "mediums in the sending of the message from its sender, and neither one of the two was a sender," Muhammad Bakhit al-Muti'i wrote. Surely the kings, princes, notables, and merchants who telegraphed each other did not take the telegraph operator to be the actual sender of the message. It therefore was irrelevant whether the telegraph operators were Muslims. Rulers appointed princes and judges via telegraphy; nobody was foolish enough to believe it was the clerk who received the telegram at the telegraph office and then typed it who was appointing these officials.[50]

• • • •

ANOTHER AUTHOR to engage with religious time by authoring a longer tract was Muhammad 'Abd al-Baqi "al-Afghani" (d. 1905). Born in Kabul, he was raised in Peshawar and migrated to Rampur in India, where he taught for a quarter century, after which he left for the Arab Peninsula and Syria to spend the remaining twenty years of his life in the Levant. Moving back and forth between Damascus and other Syrian cities, he eventually settled in Homs. In 1897, 'Abd al-Baqi published *The Book on the Useful Lessons in Opinions on the Wire and Clocks*.[51] Unlike most of the other commentators, 'Abd al-Baqi saw the timing of Ramadan to be closely related to using clocks and watches in determining the Islamic prayer times. Clocks had no immediately analogous category in Islamic law like telegraphic messages had in Prophetic hadith reports. 'Abd al-Baqi's considerations therefore instead dwelled on the topic of "innovations" (*bid'a,* an important concept in Islamic law), another crucial instrument for adapting Islam to changing circumstances.

There was a possibility that clocks had to be viewed as harmful innovations, Muhammad 'Abd al-Baqi conceded. After all, "using them and the frequency of relying on them like it is the custom in our time is a reason for avoiding to look at the sky at noon," as it was prescribed by the scriptures.[52] The five daily prayers fulfilled by observant Muslims are determined by the sun and the moon's position in the sky. Historically, prayers were measured and defined in terms of shadow lengths during the day and twilight phenomena at nighttime, all of which vary with latitude and longitude.[53] The day starts with the *maghrib* or sunset prayer, followed by the evening prayer *(ʿishāʾ)* at nightfall. *Fajr* is prayed at daybreak. Shortly after the astronomical midday when the sun has crossed the meridian is the time for *ẓuhr,* the noon prayer. *ʿAṣr,* the afternoon prayer, begins when the shadow of an object has increased over an amount equal to the length of the object casting the shadow. Each of the prayers may be performed during a limited interval of time, although an earlier prayer yields more reward.[54]

Muhammad taught the correct practice of prayer to his community and made his instructions for watching the skies with the naked eye sufficiently clear. ʿAbd al-Baqi wondered whether any justification for using clocks to aid with the prayer must not be considered an invalid pretext.[55] But after assessing the scriptural evidence about prayer times once again, he concluded that watches may be classified as an "innovation of custom," which did not necessarily mean that legally mandated practices such as monitoring the position of the sun would be abandoned in its stead.[56] For the fivefold daily prayer to be valid, Muslims have to follow certain standards of purity and practice, for instance, praying in the direction of Mecca. They are moreover obliged to know when the time for prayer comes, or at least presume to know. Relying on the *adhān* (the Muezzin's call for prayer) was akin to "knowing." So was using correct sundials and reliable ("tested") watches, as al-Afghani found. "Tested" equaled repetition, which was usually understood to increase the degree of certainty in an assumption. In this, clocks resembled shooting the cannons and lighting the candles upon the beginning of the fast-breaking holiday. Such "signs," al-Afghani stated, provided evidence that was as certain as it would get—akin to smoke pointing to the existence of fire.[57] All in all, al-Afghani found that using clocks for determining the correct prayer times could be accomplished within a shariʿa framework, and as long as clocks did not undermine practices regulated by the shariʿa, such as the observation of shadows and skylight, they did not constitute a harmful innovation.[58]

ʿAbd al-Baqi then discussed the telegram. The telegraph had become a "general custom" in handling secular messages. As a then-recent invention, the telegraph was "utterly unknown, however, from a religious legal standpoint," ʿAbd al-Baqi admitted. He then boldly declared that as far as he was concerned, the telegraph was "as reliable . . . as anything else recorded in writing." As proof, he adduced that the Prophet Muhammad himself once had a letter delivered by one of his Companions to Heraclius, the

Byzantine emperor, and Khusraw, the Sassanian emperor, calling on them to convert to Islam. Surely Muhammad would not have relied on a piece of writing if letters were deemed unreliable in Islamic law.[59] The telegraph therefore resembled the clock: both were manmade inventions, and assuming that both had been tested for accuracy, it would not make sense to allow one but not the other.[60]

• • • •

AT THE BEGINNING of the twentieth century, the lawful practice of timing Ramadan had become a contested matter in more than one way. Trustworthiness and reliability were the immediate points of contention in discussing telegraphy and its role in religious timing. Besides telegraphy, astronomic calculation of both prayer times and lunar months had come under scrutiny from learned men and legal scholars. Astronomers had long been able to calculate the rising times of the sun and the moon with considerable accuracy. From the point of view of mathematical astronomy, the new moon was considered to have risen once it reached the point of conjunction—when the sun and the moon are on opposite sides of the earth. While the point of conjunction could be determined astronomically, visibility was an altogether different matter. Visibility depended on weather conditions as well as certain qualities of the skylight that differed with latitude and longitude. Astronomers therefore were able (and are able today) to calculate the point of conjunction and to indicate the earliest time after conjunction during which the moon *could* possibly be seen in different parts of the world. But actual visibility remains contingent, at least within a certain margin.

To address Islam's preoccupation with religious time, Muslim astronomers early on developed an entire science of timekeeping, the "knowledge of the appointed times," ʿilm al-mīqāt. As a consequence, "from the middle of the ninth century until the middle of the sixteenth century it was the Islamic astronomers in the observatories of Isfahan, Maragha, and Samarqand, not those in Paris, London, or Rome, who achieved major advances in the science." Only in the late sixteenth century did European equipment, precision, and mathematics match those of Islamic astronomers.[61] As part of Islamic astronomical expertise, around the time of the thirteenth century, a professional timekeeper became a common presence in most mosques and religious schools.[62] The so-called *muwaqqit,* as his Arabic name indicated, was responsible for determining the "appointed" times of religious rituals and "constructed instruments, wrote treatises on spherical astronomy, and gave instruction to students."[63] Combined with tools such as astrolabes and quadrants, prayer tables allowed timekeepers to determine the times of the five daily prayers without being fully versed in mathematical astronomy. Muwaqqits continued to perform their profession until the early twentieth century, when gradually the spread of mechanical watches and the co-optation of religious personnel by the state spelled the end of the religious timekeeper. In many postcolonial states, new survey departments took over the task of astronomically determining times, which were passed on to religious leaders who now themselves were employees of the state.

Besides tables listing the rising times of the sun and the moon, necessarily less accurate tables existed for moon-sighting. Astronomers determined the possibility of sighting the crescent on a given day based on knowledge of the positions of the sun and the moon relative to each other and to the local horizon. They knew that the lunar crescent would be visible after sunset if it was far enough away from the sun and high enough above the horizon to not be overshadowed by sky glow in the background.[64] Since roughly the ninth century, Muslim astronomers began compiling tables showing visibility predictions for the first evening of every month. Calculations allowed astronomers to give a verdict of either "will be seen clearly," "will be seen with difficulty," or "will not be seen at all."[65] Since the Prophet Muhammad had explicitly instructed his followers to "sight" and not to calculate the new moon, the opposition of mathematical astronomy and the naked eye was not new in the nineteenth and early twentieth century. What changed was the availability of calculated tables.

The proliferation of print in the Levant not only permitted the circulation of newspapers, magazines, and reformist ideas. Print also meant that prayer and moon-sighting tables became easier and cheaper to produce than ever before. With the spread of calendars and almanacs that often included such information, knowledge about mathematically calculated religious time was available to a growing number of people. One prominent example of such a collection of tables was Ahmad Musa al-Zarqawi's *The Book of the Appointed Times* published in Cairo in 1912. Al-Zarqawi was a teacher of Islamic astronomy at al-Azhar University and dedicated the book to his students and those interested in performing basic calculations of religious duties themselves.[66] With such knowledge at hand, it was no surprise then that awareness grew of the apparent contradictions between astronomy and the human eye. On more than one occasion, people claimed to have sighted the crescent at a moment when mathematically it was not yet possible.

Such discrepancies did not go unnoticed among contemporary scholars who in turn felt pressed to take a stance in yet another important question, that of uniform practice and Muslim unity. Muhammad Rashid Rida (1865–1935), perhaps the central religious reformer, intellectual, and publicist of the period, was the editor of the widely circulating journal *al-Manar.* Rida turned *al-Manar* into a forum where he published not only his own and other authors' articles, but also where he answered questions mailed to him from Muslims as far away as Southeast Asia and Latin America who sought guidance. He offered it in the form of fatwas printed together with the original questions. Born in present-day Lebanon, Rida later left for Cairo when Ottoman censorship grew too rigid in the Levant. Like the salafis, he advocated a return to the scriptures of Islam and the practices of the Prophet and his Companions. When Jamal al-Din al-Qasimi undertook a trip of several weeks to Cairo in the winter of 1903–1904, he and his travel companion and fellow author on telegraphy, ʿAbd al-Razzaq al-Bitar, met and discussed extensively with Rida.[67]

In numerous fatwas, Rida expressed his firm conviction that the Islamic scriptures mandated "sighting," not "the calculations of the calculators and the calendars of the astronomers."[68] Islam was a "general religion" suitable for the nomadic bedouin and the sedentary urbanite alike, and it was therefore mandatory that the times of the religious duties be intelligible and known to the general public rather than exclusive to a small, elect party of calculators.[69] The higher-ups alone should never be allowed to change the rules of religious rituals and the religion according to their particularist interpretation, Rida opined.[70] He made his view even more explicit when he commented on an event in 1904. In that year, on the night of the 30th of Shaʿban (the month before Ramadan), the qadi of Egypt and members of the shariʿa court and a number of scholars gathered to hear the testimony of those seeking to sight the new moon. But not one of them testified to having seen the crescent because according to astronomical calculations, it was impossible to see the crescent at that point. Everybody with knowledge of this situation found the testimony to be impossible and considered whoever would give a testimony to be a liar. On the night of that evening, the qadi of Egypt did however receive a telegram from the judge of the town of Fayyum (south of Cairo), announcing that two witnesses had seen the crescent and that he had ruled on the beginning of Ramadan. But the higher-ranking qadi of Egypt said he could not make a decision on the basis of a telegram. However, since he had no doubt that the two witnesses in fact gave a valid testimony, despite existing astronomical calculations, he suggested they appear in front of him in Cairo and repeat their testimony. They did, and the qadi of Egypt then officially announced the beginning of the fast. For a significant number of people, the announcement came too late: they missed a day of the fast.

Many people who understood the situation were confused for two reasons. First, they had apparently followed an inaccurate calendar or other prayer table. As the testimony had been legally verified, it had to be correct. The calculation indicated in an almanac or table therefore had to be false. Second, in theory it was assumed that legally, there was no difference between the qadi of Egypt and the judge of Fayyum in terms of authority. Rida hence asked provocatively, "Has Islamic observance become dependent on a specific leader at whose hand is correct what is not correct at the hand of someone else?" The religious duty of fasting, Rida said, did not depend on an order from a political leader, and in his somewhat unusual interpretation, not even on a legal ruling.[71] To Rida, disagreements over method and false assumptions about the importance of rank in religious matters contributed to the disunity of the Muslim community. Inaccurate data compounded the problem.

Rida did not leave it at criticizing religious and political elites. Rather, he blended egalitarian arguments with calls for Muslim unity. Uniform Islamic time, however, could not be found in calculation. Numerical information in prayer tables and printed calendars was often wrong, as demonstrated by the diverging information on display across different publications.[72] Some said the month began on such and such weekday, some pointed to another day. One reason was that the astronomical day was different from the legal day, as conjunction and visibility did not coincide. According to Islamic law, the day began on the night during which the crescent could first become visible, and that might be only the second astronomical day of the month. Some of the printed tables and almanacs were unclear about which of the two criteria they applied. To Rida, the presumable unity of the calculating method was elusive.[73]

Rida elaborated his charges against error-prone calculation in yet another account, this one concerning prayer times, not calendar times. It was known, he stated, that the Muezzins in all Islamic cities relied on calculated calendars in their knowledge of the prayer times as well as clocks, rather than on what Islamic law said about shadow length or the color of the light. This change of practice had created moments of almost schizophrenic discord. Rida then recalled a trip he once took to the countryside with some of the greatest scholars of al-Azhar University.[74] There, the mufti of Egypt saw that the twilight had vanished and got ready to perform the evening prayer. But some of the scholars told him it was not yet time, that five minutes remained until prayer time—Rida was implying that these scholars must be relying on their watches and printed tables instead of following what they saw with their own eyes. The group did pray at that moment after all, but Rida reported that after the prayer, he saw them open their pocket watches and say to one another, "Now the time [of the prayer] has begun." Hence, in an example of cognitive dissonance, they prayed knowing their prayer was correct but under the impression of deviating from what was becoming the customary practice of performing the prayer according to a mechanical watch rather than observation of the skylight.[75]

Rashid Rida frequently answered correspondence that reported disagreement over methods for determining religious times in his journal *al-Manar.* In 1907, a writer from the town of Suakin in Sudan described how the members of one community were divided into three factions based on determining the beginning of Ramadan. One group followed the well-known hadith "Fast upon seeing it" and practiced sighting or, when weather prohibited visual verification, the completion of thirty days. A second group looked out for the candles to be lit, an act that occurred only after Ramadan was declared to have begun by the highest religious official of the Ottoman Empire, the shaykh al-Islam, and news had been telegraphed to the provinces.[76] A third faction followed a practice attributed to Ja'far al-Sadiq, primarily known for being counted among the shi'ite imams. This last group determined the day of the week when Ramadan would start by counting five days from the day of the week on which it commenced during the past year, a rule of thumb that may have been practical but was not considered mathematically and astronomically accurate.

In Tunisia in 1909, the uncertainty about how to determine the Islamic calendar and, in this case, the legality of the telegram created another instance of disunity. According to a reader writing to *al-Manar,* the inhabitants of certain regions were reported to have begun the fast on a Thursday after the crescent had been verified in those parts of the country. As the reader explained, however, the people of Tunis began their fast one day later because transmitting the news of the moon-sighting to the capital was delayed because "the legal rules established by the legal scholars of Tunis did not permit Muslims in these regions to rely on . . . [information that] reached them via telegraph or telephone about testifying on the crescents of the fast and breaking the fast because the telegraph is in the hand of non-Muslims."[77] Rashid Rida's answer once more stressed the need to return to the simplicity of the basic tenets of Islamic faith. "What we have to say about the matter is easy," Rida began. "Only that the majority of Muslims has come to no longer appreciate simplicity and brevity in religion although it belongs to the origins of Islam in the text of the book [the Qur'an]."[78] In his view, the telegraph enhanced such simplicity and thus unity. Arguing against calculation and for sighting and telegraphic transmission, Rida called attention to simplicity as the source of Muslim strength.

Several years later, in 1922, Rashid Rida received another query that complained about the lack of uniformity in timing practices. The author was 'Abd Allah Ibn Muhammad al-Mas'udi, who was writing from Siam from the city of Bangkok. "Every year, differences and disunity occur among the imams of the mosques about the verification of the crescent," al-Masʿudi deplored. Some of these imams relied on prayer tables; others followed the aforementioned rule proposed by Jaʿfar al-Sadiq, which suggested counting five weekdays from the weekday on which Ramadan had started the previous year. Another group followed established sighting practices based on the "fast upon seeing it" hadith. All in all, "every mosque began fasting based on what its imam opined." Rida yet again strongly advocated sighting over any form of calculation in order to overcome such discord.[79] Later he wrote that the religious times "depended on matters witnessed with the senses so that Muslims shall not be divided and will not be in need of leaders and scholars for [determining] the times of their religion."[80]

On another occasion, Rida's journal *al-Manar* received a query from a Muslim located in the Russian Empire. Radaʾ al-Din was a judge at the shariʿa court in Ufa in what is currently the Russian Republic of Bashkortostan between the Volga and the Ural mountains, home to the Bashkir ethnic group of primarily Sunni Muslims. The judge wrote to Rida citing the "fast upon seeing it" hadith for Ramadan but wanted to know how to derive the starting dates for the months other than Ramadan and Shawwal, both of which were related to religious festivities and were therefore subjected to special rules. After all, Radaʾ al-Din found, widely available printed calendars nowadays contained all sorts of information that had been accumulated and published independent of crescent-sighting and court rulings by people with appropriate skills and knowledge. As inhabitants of the northern hemisphere, sighting the new moon at the right time was rarely possible for Muslims in the Russian Empire, Radaʾ al-Din wrote. As a consequence, the scholar from Russia feared that disunity among Muslims would only increase if the same challenging rules for sighting applied to all months, not just Ramadan and Shawwal.[81]

Radaʾ al-Din stated that his concerns had been exacerbated. Divergent views about when to start and end fasting had proliferated, and such differences "had become the laughing stock among the people of the other religious communities who live among us. Often, people in one place fast and people in another place break the fast." The spatial difference between such two places was mostly negligible with possibly even two imams at the same mosque or members of one family disagreeing.[82] Rida replied by emphasizing the need for uniformity of practice since it was possible for Muslims in all regions to agree on the times of religious rituals, as he found. He himself did not even see much discord in religious times among people of neighboring regions except when someone lied about sighting the moon or testified based on some form of imagination. What really compromised unity was that "the calendars printed in Egypt [and by implication, elsewhere] every year differed in verifying these months."[83]

Rida's misgivings about erroneous calculations were not entirely unfounded, although it was prayer times, not calendar times, that caused the greatest confusion. The Egyptian state itself finally intervened and attempted to put an end to the multitude of false printed clock times that filled columns and pages in the various almanacs on sale throughout cities like Cairo and Alexandria. In 1923, the Egyptian Ministry of Pious Endowments issued an ordinance to all heads of mosques in Cairo. The ministry obliged all those performing the call to prayer or initiating the prayer to set their watches to twelve o'clock at the firing of the noon cannon and to observe the prayer times according "to the calculation of the government almanac."[84]

The background to this was that the several independent, private almanac makers had failed to change their calculations when the Survey Department, cajoled by British astronomers at the Helwan Observatory just outside Cairo, had introduced standard time to Egypt in 1902. To distribute the new time, the department had instructed the existing noon gun to be now fired at UTC+2. As local time and UTC+2 were five minutes apart, the noon gun was now fired five minutes later than previously when it had followed local time. Muslims, however, prayed according to local sun time, not standard mean time. But in calculating prayer times, almanac makers failed to register the time change for the noon gun and continued to take the sound of the cannon as the advent of twelve o'clock local time—when what was announced was really twelve o'clock plus five minutes. Hence, almanac makers set their watches to 12:00 p.m. local time upon hearing the cannon when it was actually 12:05 local time. Since the calculators' inattention to this detail caused "agitation among the masses and confusion in the determination of the prayer times," the Ministry in 1923 moved to unify prayer times under its aegis and to make government-issued almanacs the only valid standard.[85] Such a step reflected the state's growing resolve to regulate religious affairs down to Islamic timing practices.

Rida the scholar believed that the edifice of elaborate and sophisticated doctrines of legal reasoning, piled up by erudite elites like him over the several centuries that had passed since the death of Muhammad, obfuscated the true meaning of the religion. Common Muslims stood uncomprehending in front of the muddle of technical and often hair-splittingly complicated arguments and provisions. Instead of seeking to follow their distortions, Rida urged a return to a simpler Islam that would refrain from legal speculation and the overuse of reasoning.[86] In Rida's eyes, Muslim unity would therefore spring not from following complicated (and often erroneous) calculations of the Islamic calendar but rather from relying on the sensual faculties that all humans had been equipped with. Uniformity and unanimity of Islamic time and the community of all Muslims ('umma) were to be found in the accessibility of religious rituals. Standardization of Islamic calendar time should be mandated to unclutter Islam and to return to the uniformity of simplicity.[87]

• • • •

As these debates about Islamic time and space indicate, questions of determining the times of Ramadan with the help of the telegraph had already triggered a reconceptualization of spatial categories of distance and location, interconnectedness and isolation. Telegraphy itself seemingly bridged distances between Muslim communities that only a few decades earlier would have made the timing of Ramadan a local affair. Such a reconfiguration of time and space attained particular salience in the disagreement over the so-called Difference in rising points/horizons. This aspect of the calendar controversy focused on whether locations with different longitudes should observe different start dates for Ramadan and the fast-breaking holiday, given that astronomers knew the visibility of the moon to differ according to the local horizon. If the new moon was verified in one community and a telegram informing about that community's ruling on the start of Ramadan was sent to another, how far away could the second community be, given the "difference in horizons"? And how far was far?

'Abd al-Baqi "al-Afghani," author of the book on clocks and prayer times, declared that he was unmoved by the "difference in horizons" argument, for "if it [the crescent] is verified in the West it is verified in the East."[88] Another scholar stated, there was no reference in the key religious texts to the problem of different horizons except when the distance between places was very far, "like Khurasan [in present-day Iran] from Andalusia."[89] Jamal al-Din al-Qasimi, on the other hand, ventured that "everybody who knows a piece of the science of constellation and the science of times and of geography" understood that countries with differing horizons did not share a common starting time for Ramadan. Every greater region had its own ruling.[90] As Rashid Rida detailed, some scholars said that it was not necessary for the people of one country to act based on a sighting by the people of another; others held that if the countries were located close to one another their ruling was one and the same; if they were far apart, each country acted according to its sighting. Rida himself was unwilling to dismiss a valid ruling due to astronomical arguments about different rising points: if the crescent had been properly verified in one community and a nearby location had sighted it at a different time, there was no reason to throw out the first ruling simply because theoretically, proximity and the astronomy that followed from it would have the two be identical.[91]

On the other hand, in countries in which there was "mixing" and cooperation between regions, as was the case in Egypt, it was illogical that some of the residents of these countries went into fasting while others broke the fast. However, for countries with no strong ties between them and no connections except for migration, the timing of Ramadan could only be local.[92] Overall, there was therefore a diversity of opinions among the generation of scholars who had achieved prominence around 1900 when it came to the question of longitudinal variance. Yet regardless of whether a scholar came out in favor of or opposed to taking geography and astronomy into account, the discussions about distance and telegraphic transmission raised new questions of belonging and affiliation that had not posed themselves in this form in a previous age.

It was only in the 1920s and 1930s that a new generation of Muslim scholars took the question of standardizing Islamic calendar time across vast spaces a step further. The previous generation had gradually come to endorse telegraphy as a trustworthy and reliable means of quickly relaying the news of the crescent-sighting, thus signaling the universality of Islam and the shariʿa that united Muslims throughout centuries and across the world. Yet they had shied away from firmly approving calculation. Even in the "difference of rising points/horizons" matter, a variety of opinions coexisted. Now, twenty or thirty years later, the uniformity that Rida saw in the simple, scriptualist observance of moon-sighting was no longer deemed sufficient. A Muslim community that had witnessed the end of the Ottoman Empire and the abolition of the caliphate in 1923 and 1924, and the establishment of European colonial or later euphemistically, "Mandate" states in Egypt, the Levant, and Iraq, was in more desperate need to unite than at any time during the decades preceding World War I. This was the hour for calculation to be hailed as the solution that would see Muslims all over the world rely on the same astronomic practice for determining the Islamic calendar and the start and end of Ramadan, the panacea that would help fix Muslim disunity. Not only would calculation replace sighting, it would also disregard variations in latitude and longitude, thus casting aside the "difference in horizons" argument. Unsurprisingly, such calls for reform met with criticism. But ever since the late nineteenth and early twentieth century, schemes for uniting the religious times of Muslims all over the world surfaced with some regularity even to this day.

In this case, it was not the government that sought to clamp down on temporal irregularity as in the case of Egyptian prayer times, but a member of the cohort of Muslim learned men. Likely the first to call for a universal

Islamic calendar was a scholar named Ahmad Shakir (1892–1958). Born in Cairo to a family of distinguished learned men, he himself became a renowned expert on prophetic reports and commenter on the Qur'an as well as a jurist. Once he enrolled at al-Azhar University, Shakir studied with known reformers Rashid Rida and Taha al-Jaza'iri. But like others who advocated reforms in one field—in this case, Islamic timekeeping—Ahmad Shakir strongly opposed other forms of change and fervently attacked Westernization in other areas. Shakir later rose through the ranks of the judiciary and eventually became chief justice at the shari'a high court of Egypt.[93] Ahmad Shakir admitted that he had only gradually warmed to the idea of practicing calculation instead of sighting. Not long ago, he had strongly opposed to a proposal for a minor change in crescent-sighting rules: another well-known Azhar scholar and reformer, Muhammad Ibn Mustafa al-Maraghi had proposed that no evidence of naked-eye crescent-sighting should be accepted if scientific knowledge and thus, calculation, contradicted such a testimony. Al-Maraghi's fatwa stirred up yet another heated controversy and was rejected by the majority of influential scholars.[94]

In 1939, Shakir authored a tract that went one step further. Titled somewhat innocently *Beginnings of the Arabic Months,* the essay laid out Shakir's plans for adopting a unified, scientifically based Islamic lunar calendar. The calculated calendar would avoid the pitfalls of instability and confusion that arose from determining the beginning of every month (and Ramadan as well as the time of the Muslim pilgrimage to Mecca, the Hajj in particular) by sighting rather than calculation. As in [illegible] other instances, Shakir's call for reform began with an account of [illegible] egregious example of heterogeneous religious time. In 19[illegible] shari'a court in Egypt established that the month of Dh[illegible] on Saturday, January 20, hence the Festival of Sacrifice ([illegible] fall on January 30. A few days later, a magazine repo[illegible] Arabian government had declared that not Saturday, [illegible] was the first of the month. The holiday therefore fell o[illegible] Later that month, the Bombay correspondent of anot[illegible] ported that the Muslims of the West Indian metropolis had c[illegible] on Wednesday, implying that in India, the first of the month of Dhu al-[illegible] was Monday.[95] Moreover, in the year Shakir was writing, another debate was raging about determining the day of *'arafa,* known to Muslims as the supreme day of the pilgrimage season.[96]

Ahmad Shakir was keenly aware of the previous generations' deliberations on using the telegraph and the telephone in relaying crescent-sightings from one part of the world to another. "Muslim countries have virtually all become one in being able to receive messages from anywhere in the world," he stated. Only a year earlier, a detailed inquiry from India had been sent to the senior scholars of al-Azhar University asking for guidance on the question of stabilizing the beginning of the months. The request was subsequently circulated among a leading group of scholars, one of whom was Ahmad Shakir's father. Shakir took the existence of such far-flung ties of communication and exchange among Muslims as another token of the increasingly global nature of Islam. Ahmad Shakir admitted that the variance of horizons had always been a contested matter among the scholars and different legal traditions of Islam. Men of learning differed on what constituted "far" and "near" countries and whether generally the visibility of the crescent in one country had legal consequences for the beginning of the month in other countries. Did every country have its own horizon?

Shakir's arguments against strictly observing the "difference in horizons" were similar to those advanced by clock time reformers such as Sandford Fleming for converting local times into mean times. Ahmad Shakir pointed out that taken to its logical conclusion, observing the difference in horizons meant that strictly speaking, every locality would have its own month or calendar since it had its own ever so slightly different crescent visibility. Earlier Euro-American discussions about the adoption of mean times and the conditions of each locality following its own time had irked the advocates of standard time on the American continent and in Europe alike. Shakir and likeminded Muslim calendar reformers eschewed reverting to a concept of "mean" calendar times to be substituted for the plurality of local times. The "true month," Shakir explained, astronomically began with conjunction. And since there was only one moon that rose, there was but one single calendar for all Muslims.[97]

In advertising calculation, Shakir focused on the differences between the needs and skills of early Muslims and the 'umma in the 1930s: in a largely tribal and initially widely illiterate setting, observing the heavenly bodies, shadows, and skylights was the only adequate method of timekeeping. Now that a majority of Muslims was literate according to Shakir, it was not merely permissible but mandatory for Muslims to rely on more stable and certain methods of determining religious times. Sighting ought to be reserved to those moments when remote rural areas and villages were unable to procure authentic and accurate information by expert calculators.[98] Calculation increased certainty, Shakir noted with a nod to Islamic legal theory, and attaining the utmost degree of certainty in the performance of religious obligations was a duty. Rules changed when the conditions changed, Shakir observed, not unlike Jamal al-Din al-Qasimi, who had similarly praised the adaptability of Islam.

Ahmad Shakir recommended that Mecca be chosen as the universal point of reference for calculating the moment of conjunction. As a reason for this choice he gave an unorthodox and complicated reading of classical texts that, according to Shakir, proved that historically, the Muslim community had begun to fast when the people of Mecca fasted. Hence in the modern world, Mecca would come to be host to an Islamic time institute that would scientifically establish the time and date for all believers. "Muslims will be united in identifying the lunar months," and Mecca, "the fountainhead and cradle of Islam," would be the center of Islamic timekeeping.[99] In some ways, Ahmad Shakir's choice of an Islamic institution in charge of timekeeping had been anticipated by Rashid Rida roughly thirty years earlier. In 1903, Rida contemplated that if one day it became feasible for all world regions to communicate with one another via the telegraph, and if the Muslims had a Great Imam (which was then still the Ottoman caliph) whose rulings were executed across all their lands, and if that leader could easily notify all Muslims by telegraph about legally verified crescent-sightings, the community of all Muslims, the global 'umma, stood to benefit immensely.[100] Rida's broad vision of Ottoman-led Muslim unity was not to last. With the dissolution of the Ottoman Empire and the abolition of the Caliphate, Islamic internationalism lost its older pan-Islamic gloss and turned into an increasingly Saudi Arabian brand of ossified, conservative religion whose leaders would have scorned the mere idea of using modern technology in determining Islamic calendar time.[101]

Ahmad Shakir's proposal was met with opposition from conservative circles as soon as it became public, but his critics failed to silence calls for Muslim calendar uniformity in the long run.[102] In the latter half of the twentieth century, the unpredictable and nonstandardized nature of the lunar calendar found its way onto the agendas of several international bodies. In 1955, Jordan urged the Arab League to address the question of Islamic holidays. In 1961, the League sent a request to al-Azhar University in which it asked the scholars to investigate in principle a uniform lunar calendar and prayer timetable. International bodies like the Saudi-based Islamic Fiqh Academy (affiliated with the Organization of the Islamic Conference/OIC, the highest international Islamic organization) have also dealt with the question of a uniform calendar. Yet the conservative Saudi-dominated Islamic Fiqh Academy only conceded that astronomical calculations may assist the mandatory sighting of the crescent.[103] In the second half of the twentieth century, migration from the Muslim world to Europe and North America has only compounded the dispute over whom to follow in fasting. Despite these initiatives, a consensus on Islamic calendar unification therefore remains elusive.

As the wide range of interventions in the debate about Islamic calendar and clock times illustrates, this was and is a topic of persisting interest and concern among scholars, the public, and government officials alike. When Muhammad Rashid Rida and others found themselves in the company of other scholars and legal figures, the conversation habitually turned to telegraphic messages, calculation, and the determination of Ramadan. Newspapers and journals helped contemporaries keep up with these debates as they reported on irregularities and disagreements in the timing of the holy month and the fast-breaking holiday. Methodologically, analogies drawn by scholars, jurists, and judges between telegrams and reports about the Prophet's life are a textbook case of how reformist thought worked, and how Islam adapted to a rapidly changing technological and political environment.

As in Europe and North America, print, telegraphy, and interconnectedness led Muslims to experience time, space, and belonging in new ways. In turn, like their Euro-American contemporaries, Muslims questioned conventional notions of community. Using the telegraph to transmit the news of the moon-sighting acutely raised the question of just how far communities could be apart to still observe the same lunar months and related religious rituals. Telegraphy initially increased divergent determinations of calendar time, as doubts about the legality of telegraphy prevailed. What is more, the use of the telegraph first and foremost brought evidence of dissimilar calendar timing into the full light of day and turned the matter into a concern and issue in the first place. In an interconnected world, differences came into sharp relief. Once endorsed by legal scholars and learned men, however, telegraphy was praised as a means to overcome not only differing opinions on the legality of the telegraph, but a host of other varying practices of calendar determination as well.

"Print capitalism," the serialized daily newspaper that displayed the same date on its header to people scattered all over different locales, allowed readers to imagine themselves as members of national communities. Calendars and almanacs, clocks and telegraphy, can be viewed as part of a similarly "industrial," serialized time.[104] Yet contrary to what Benedict Anderson posited for Europe, imagined communities were not national alone. In Europe and North America, transformations of time and space inspired scientists, railway men, and other elite reformers to picture a world that was turning into a smaller place. Their image of the world that was interconnected, in which travelers moved with ease and speed, was based primarily on Europe and North America. When Europeans talked "interconnectedness" in describing the annihilation of time and space, or heralded internationalist institutions and agreements as part of international society, they implicitly envisioned a community of Europeans, North Americans, and possibly their offspring in Australia under exclusion of other world regions and societies.

Similar essentialized reconfigurations of global space emerged in conjunction with nineteenth-century globalization outside the West. Certainly in the Islamic world of the Middle East but even far beyond it in the Russian Empire and Southeast Asia, telegraphy, print products, and a changing experience of time and space inspired late nineteenth- and early twentieth-century Muslims to rethink the extent of the Muslim community or 'umma not as national but as global. Pan-Asianism and Pan-Africanism bore similar traits. It was the prerogative of Euro-American visions of an interconnected world that such worldviews were phrased in universalizing claims and disguised behind the all-inclusive language of "world" standard time and, as we shall see, a "world" calendar.[105]

Muslim thinkers ultimately proposed a calendar internationalism that resembled Euro-American movements of the same period, even though a universal Islamic calendar eschewed the all-comprehensive notion of "world." Still, the function of such a universalization of Islamic calendar time, the categories and terms that linked it to Islamic law and the Islamic tradition, cannot be said to have emulated Euro-American norms and concepts. Muslims debated the mechanization of Islamic calendar times through telegraphy and the standardization of calendar determination via calculation for their own purposes, on and in their own terms.[106] And yet, while Ahmad Shakir was gradually coming to accept and promote the idea of a global Islamic calendar, officials roaming the halls of Geneva's splendid League of Nations Palace weighed the introduction of a uniform world calendar to replace calendars specific to particular religions and societies around the globe. The world calendar activists at the League of Nations and beyond and the Muslim calendar reformers responded to the same overarching political, economic, and social development of an increasingly interconnected world. But while similar in form (a uniform calendar of global reach), the specific function and purpose of Euro-American calendar reform proposals and the motives behind them were significantly different from Muslim efforts. Simultaneity did not always entail identity. Indeed a general similarity of interests and preoccupations did not even necessarily emerge from the flow of ideas and concepts from one part of the world to another. In some instances, as was the case in Britain's India, in France, and in Germany, local identities and the particularities of a political situation shaped highly particular appropriations of universal time. But globalization could also inspire surprisingly similar reflections on time originating in different parts of the world in widely different societies.

CHAPTER SEVEN

# One Calendar for All

ONE DAY in 1884, the French astronomer Camille Flammarion received a visitor. It was the same year in which the conference in Washington, DC, would pass resolutions about the adoption of a prime meridian, and Flammarion and other astronomers were likely well informed about these goings-on for they avidly followed the scientific and popular-scientific press. Flammarion, whose brother would establish the well-known French publishing house of the same name, was the founder of France's first astronomic society and publisher of the leading French journal in that field.[1] Flammarion's visitor was a prison chaplain who had just returned from a trip to Rome where he spoke to the Vatican about reforming the Gregorian calendar by stabilizing the date of Easter. The Holy See ordered the chaplain to return to Paris and seek the feedback of well-known astronomers, eventually leading the chaplain to knock on Flammarion's door. During their conversation, the chaplain suddenly produced 5,000 Francs from his pocket, "received from an anonymous donor" eager to endow a prize for the reform of the calendar.[2]

It took three more visits to convince Flammarion to take the money, and a few months for the project to come to fruition, but in the fall of 1884, just one month before the Washington conference opened, Camille Flammarion's astronomical journal published a call for submissions of calendar reform plans.[3] The criteria for reform calendars were specified as follows: the new calendar had to permit every date to always fall on the same day of the week, and the length of months had to be as even as possible. Flammarion together with three colleagues adjudicated the submissions. Based on the stipulations set forth by Flammarion and the anonymous donor, the six prized proposals, announced in 1887, detailed a format that would likewise be at the heart of calendar discussions over the coming decades.[4]

The French calendar competition was not the first mention of calendar reform in the late nineteenth century. Yet it is of particular interest due to the close temporal and personnel connections it reveals between calendar and clock time reform as both were unfolding over the very same years. Henri Poincaré, the famed mathematician and physicist whose work has been associated with the adoption of mean times and the telegraphic synchronization of clocks, presided over a later session of the Astronomical Society where Flammarion revisited the history of the calendar reform competition; at the same session, it was reported that the French legislator who had drafted one of the bills proposing to adopt universal time in France in the late 1890s was now engaged in talks with the Vatican about stabilizing Easter and reforming the calendar.[5]

In later decades, MP Robert Pearce, sponsor of the daylight saving bill in Britain in 1908, brought a calendar reform bill before the House in the same year. In Germany, Wilhelm Förster, head of the Berlin Observatory and a leading voice in German debates about Central European Time, simultaneously provided expert advice to the German government on calendar reform and authored several publications on the topic.[6] Once calendar reform made it before the League of Nations in the 1920s, Guillaume Bigourdan, a central figure in French astronomy who was deeply involved with the adoption of French mean times and time distribution in the 1880s, 1890s, and 1910s, was appointed as a member of the League of Nations' committee for the study of calendar reform.[7] In the 1950s, Harold Spencer Jones, then Astronomer Royal at the Greenwich Observatory, still gave talks about the reform of the calendar.[8]

There was even a material side to the intertwined nature of clock and calendar time: one of the latest fashions among clockmakers in the late nineteenth century was so-called calendar clocks, advertised as office and school supplies. On their several dials, such clocks indicated not only hours and minutes but also days, months, and the year.[9] Captured by the vision of overcoming the difference in calendars around the world, scientists and clockmakers were joined by legislators and administrators in their engagement with calendar reform. Prior to the outbreak of World War I, three European parliaments discussed the stabilization of Easter and a more general reform of the calendar. In light of these close connections between clock and calendar time, then, it is all the more surprising that attempts to unify calendars are entirely absent from existing histories of the science behind time reform in the later nineteenth and early twentieth century, for contemporaries clearly viewed the unification of clock times and calendars as pertaining to the same overarching question of reforming "time."

More than in the case of clock times, economic and capitalist interests enthusiastically rallied behind calendar reform. In addition to scientists and the usual slew of self-styled reformers and government officials, business owners, economists, and statisticians were the most visible actors in promoting a uniform calendar. It was here that a connection between capitalism and homogenous time emerged, rather than in the adoption of mean times. Yet this relationship was unlike that assumed by E. P. Thompson and those following in his footsteps. Uniform units of time came to matter to economists and capitalists not for the inculcation of time discipline but in the context of producing economic knowledge and rendering countries economically legible in the manner that Timothy Mitchell has analyzed for Egypt in his *Rule of Experts*.[10]

Overall, calendar reform was arguably the aspect of time unification that resonated most widely while simultaneously stirring up the most heated controversies. Countless brochures and other publications devoted to the topic, a whirlwind of publicity unleashed by the League of Nations' activities and its similarly well-organized opponents in different church and religious circles, the lobbyism of several private international nongovernmental organizations (NGOs), and economic and business preferences all testify to the scope of interest in calendar time. Once governments discovered their taste for calendar reform, however, universal calendar time quickly acquired national, and nationalist, meanings and functions. The reform of calendars attracted arguably even more attention in a burgeoning international public sphere than the unification of clock times ever had. But not least because calendar reform failed in the long run and because the unification of clock times was mostly accomplished by the middle decades of the twentieth century, historians of science have overlooked the importance of calendar reform over several decades.

Calendars and dating systems were another tool for positioning societies and nations in historical time by inspiring comparison and conversion between different systems. Calendar time, and the historical and religious implications it bore, compelled contemporaries to understand these histories in relation to others. Calendar reform was much discussed beyond Europe and North America. Especially in the years surrounding a number of revolutions in the non-Western world at the beginning of the twentieth century, calendar reform took center stage: China at least partially and nominally introduced the Gregorian calendar as of 1912 in the context of the Chinese Revolution a year earlier. Following the Young Turk Revolution of 1908, a plan was placed before the reinstated Ottoman parliament to adopt Greenwich time for certain purposes, although it failed to pass for the time being.[11] Following the Constitutional Revolution in Iran in 1906, attempts were made to further the use of the solar year in fiscal matters, even if Iran's current solar calendar was only formally introduced in 1925 with the ascension of a new dynasty. Calendar reform is a stark reminder of the simultaneous and global nature of efforts to unify and improve time measurements. Euro-American designs for a uniform world calendar are only one manifestation of this wider trend.

The story of the enterprising men and women who labored to make calendar time uniform substantiates the importance of individuals in global history. International organizations like the League of Nations through their cooperation with NGOs opened up new opportunities for men and increasingly women to engage in lower-level diplomacy. Geneva in the interwar years was where the cast of ordinary political actors turned more diverse in many ways. Such autodidactic diplomats served as mediators between the world of international organizations, national governments, and regional civil society associations. Calendar reform relied on women like Elisabeth Achelis, founder of the World Calendar Organization, who brokered between diplomatic circles, League of Nations personnel, and countless "reform" clubs and associations from the United States to South Asia.

• • • •

CALENDAR REFORM emerged as a two-pronged project. Early on in the 1870s and 1880s, it often consisted of plans to urge Russia and other Orthodox societies to adopt the Gregorian calendar. From the 1880s and 1890s, such ideas targeting Orthodoxy continued to be voiced occasionally, but overall, the debate shifted to improving the Gregorian calendar and rendering it universal.[12] One of the most vocal early activists of extending the Gregorian calendar into Russia was Cesario Tondini de Quarenghi. A Barnabite monk and promoter of Catholicism, he spent several years in Paris in the 1860s and 1870s, where he established contacts with Russian expatriates. Back in Italy, he succeeded in convincing the Academy of Sciences in Bologna to support his plans. During the close sequence of internationalist meetings discussing time unification in the 1880s, the Bologna Academy repeatedly published material promoting the meridian of Jerusalem as the world's base meridian. Behind these ideas was usually Tondini de Quarenghi. In the same breath, the Italian tied his advocacy for uniform clock times to what he and others painted as the "unification" of calendars.[13]

In the 1880s, Tondini de Quarenghi was writing to the Porte in Istanbul, asking the Ottoman government to officially follow the Gregorian calendar. Reports also placed him in Sofia and eventually in Russia. There, he managed to get himself expelled due to the missionary zeal he displayed in his interactions with Orthodox clerics and the imperial government, for "he entirely lacks a sense of established diplomatic forms," as Wilhelm Förster, the German astronomer, once described him. In 1888, now residing in London, he could be found at the annual meeting of the British Association for the Advancement of Science, giving a paper on Russia's attitude toward calendar reform. A new generation of self-taught "ambassadors" emerging in interwar Geneva would grow more skilled at navigating the waters of international activism, but Tondini de Quarenghi prefigured the emergence of a corps of unofficial but influential diplomat-activists.[14]

The fixation on Russia may have been Tondini de Quarenghi's issue in particular, but he was not alone with his interest in the times of the Russian east. Russian authorities themselves began to consider some aspects of the Orthodox calendar to cause problems. With the new millennium and the caesura of 1900 approaching, talk of adopting the Gregorian calendar in Russia grew louder, as the imperial government and the Academy of Sciences in St. Petersburg now considered such a move. In 1910, the Tsarist government was reported to have completed an investigation into the unusually high number of holidays celebrated in Orthodox Christianity. In some districts, the number of holidays exceeded the number of workdays in a year; in others, holidays still made up about one-third of all days. During the main harvesting months of May, June, and July, the frequency of holidays was particularly difficult to sustain. Customs varied starkly among districts and towns. A number of government holidays were observed by some though not all official institutions.[15] But the Russian government never wholeheartedly pursued calendar reform, and most importantly, different Orthodox churches remained divided over the question. By the outbreak of World War I, Orthodox Christianity had weighed in repeatedly on whether to adopt the Gregorian calendar, with most still opposing such a step. Orthodox clergy declared their consent to be dependent on Catholics and Protestants ceasing to "fight" Orthodoxy; relations between the Christian churches were fraught.[16]

In this atmosphere, the terms of the calendar reform debate shifted. Commercial ties to the Orthodox east no longer galvanized attention. Calendar reform veered away from Russia and Orthodoxy to the stabilization of the Easter holiday and uniform months and quarters in the Gregorian calendar. Worldwide comparability and stability were now the goal. As one German newspaper held forth, "the ongoing internationalization of civilized life makes palpable in ever increasing degrees the formal differences that nowadays separate different civilizations"—it was interconnectedness, as Nietzsche had concluded, that rendered differences visible. "Formal international uniformity" was a requirement in many aspects of life, and one area of such a quest for uniformity was that of calendar times.[17] As if leafing through a Beirut-issued almanac, contemporaries were aware that calendar pluralism was exacerbated in locales where various communities cohabited. The article explained, people living in the Levant normally had to operate all of their different calendars simultaneously, and occasionally even an additional Coptic calendar. In Poland, Greek Catholic, Roman Catholic, Protestant, and Jewish holidays were celebrated alongside each other, resulting in a reduced number of commonly shared working days. But contrary to Arab observers who converted and calculated back and forth between different times, European and American calendar reformers dreamed of a worldwide uniform timekeeper as the solution to such heterogeneity.[18]

Neither the leap year solution nor the gradual accumulation of divergence between the solar year and even the "new" reformed Gregorian calendar were tagged for reform. What was now seen as the most disadvantageous element in the calendar was its "irregular" subsections—the fact that months contained between twenty-eight and thirty-one days, that there was no systematic rule behind this allocation of numbers of days to certain months, and that dates fell on different weekdays every year. Easter, which could fall as early as March 25 and as late as April 22, was another target for more stability and uniformity.[19] Some of these points had been raised as early as the French competition arranged by Camille Flammarion at the behest of the anonymous donor; by the turn of the century, they stood front and center. A few overly radical reform schemes would always surface but soon prove too drastic to be seriously considered, for they usually involved a wholesale abandonment of the unit of the year or month. A yogi from northern India wrote more than two-dozen letters to the Astronomer Royal at Greenwich in the 1950s, intent on convincing him of his "Improved Rhythmic Calendar" of Indian yogic origin.[20] Proposals garnering more serious interest largely fell into two groups.

The first set of formats worked off a design popularized by the French positivist philosopher Auguste Comte at the beginning of the nineteenth century, although in most cases the advocates of this scheme did not make explicit connections to Comte's calendar and may not have been aware of the precedent. Comte's positivist calendar followed a year of thirteen months. In the late nineteenth and early twentieth century, such thirteen-month plans gained considerable popularity, as they offered the most even month length of all with twenty-eight days each. The second group of reform schemes normally eschewed the more drastic intervention of adding a thirteenth month and instead suggested to reduce the number of days in the year to 364, divisible into four even quarters of ninety-one days, while adding an additional "blank day" that remained outside the count of the week. The intercalated blank day ensured that dates would always fall on the same weekday (January 1, for instance, could be made to always fall on a Sunday), thus rendering the calendar perpetual. The latter variant had been popularized by a professor of mathematics at the University of Geneva in the 1880s and 1890, L. A. Grosclaude, and had found supporters in France, Belgium, and Britain.[21] In 1895 one Ignatz Heising, an engineer and German immigrant in Pittsburg, PA, wrote to the German Ministry of Foreign Affairs to make known his plans for a new "era" for the entire world, by adopting a perpetual calendar strongly resembling Grosclaude's.[22] Some of the self-declared reformers moving in more scientific-minded circles were well aware of other activities in the field through journals and other publications. But more often, individuals scattered across the globe simultaneously came up with strikingly similar answers to the "problem" of irregular units of calendar time.

• • • •

SETTING THE EARLY PARAMETERS for discussing calendar time was not least the achievement of organized business interests. In the later nineteenth century, a number of national Chambers of Commerce from various countries had formed an international association, devoted to fighting protectionist politics that had become so prevalent during these decades, while at the same time furthering national economic interests.[23] When the international association of Chambers of Commerce met in Milan for one of its first congresses in 1906, it discussed the economic implications of uneven quarters and month lengths and the date of Easter. Three further meetings of the Chambers in Prague in 1908, London in 1910, and Boston in 1912 took up the question of the calendar anew.[24]

The flurry of publications and meetings addressing calendar reform soon captured the imagination of national legislators. The Chambers of Commerce now organized a concerted lobbying effort to get governments on board, and the Swiss government agreed to discuss the possibility of a conference on calendar reform and Easter in the Swiss upper house. Everybody realized that securing the Vatican's attendance at such an event was crucial. Since Switzerland lacked diplomatic representation at the Vatican, it instead convinced Belgium, a largely Catholic nation, to step into the breach. Two years later, in 1912, it was all too clear that the Vatican would not consent to a conference, and the move for an international congress faded. But the interest of legislators and governments had been piqued.[25]

After the Swiss government had reached out to other European authorities about a conference, the British Board of Trade began investigating the impact of changing the calendar in 1908. That same year, the ink on his daylight saving bill having barely dried, British MP Robert Pearce sponsored a bill that proposed both to stabilize Easter once and for all and to amend the Gregorian calendar in such a way that would absolve it of its harmful irregularities.[26] Pearce's scheme entailed a blank day that would be intercalated between December 31 and January 1, called New Year's Day, without counting as a regular weekday. The remaining days would form four quarters of 91 days each, consisting of three months of 30, 30, and 31 days. Easter Sunday would forever be fixed to fall on Sunday, April 7. His efforts remained unrequited for the time being. The British government withheld its support for Pearce's bill and again when slightly modified calendar reform bills were introduced in 1911 and 1912.[27] In Germany, the Protestant Church declared itself in favor of some measure to stabilize Easter, instilling hope in government officials that calendar reform and the Easter question might stand a chance of acceptance. When citizens submitted petitions to the German Parliament asking it to consider the calendar question, however, these materials were passed on to higher ranks for information but no action was taken either. From the onset, calendar reform was understood to be a more touchy affair than the unification of clock times had been due to a mix of religious and nationalist unease.[28]

World War I brought a temporary end to calendar reform. In the early 1920s, plans for reforming the Gregorian calendar were reanimated and placed under new auspices. The Chambers of Commerce realized it would require the cumulative power of national governments combined and the international legitimacy provided by the League of Nations to further reform. Again relying on the help of the Swiss government, the Chambers managed to convince the League of Nations to study the possibility of organizing a congress on calendar reform in order to gather the "ecclesiastical, scientific, and business world" around the topic.[29] The question of reforming the calendar thus came before the League's Communications and Transit Section, where it was decided to collect the necessary documentation to assess the needs and chances for calendar reform. The League of Nations, set up in Geneva as part of the Treaty of Versailles after the war, has been shown to have functioned above all as an international clearinghouse, a sizable bureaucratic machinery gathering information and documentation and publicizing its findings and concerns with a keen eye on public opinion. Calendar reform supports this view.[30]

Geneva's method of assembling information on calendars and other causes it underwrote consisted in mailing off questionnaires to its liaisons in different member countries and asking members to conduct their own surveys of public opinion on calendars. The League of Nations' first report on calendar reform, published in 1922 based on the responses from governments and associations, listed a number of flaws in present calendar arrangements that warranted reform: the date and day of the week did not correspond from one year to the next; months were of diverging length; Western and Eastern Europe, Roman Catholic and Orthodox countries, used different calendars. In light of these shortcomings, League officials expressed their support for reform and their intention to devote further study to the calendar question, as long as reform was "definitely demanded by public opinion with a view to an improvement of public life and economic relations." From the outset, the League shunned taking on religious interests. It declared that questions such as the stabilization of Easter ought best to be left to religious authorities and that generally, it was not "incumbent" upon a League of Nations committee to express any opinion as to which particular calendar reform format was the most recommendable.[31]

Members of the League's administrative apparatus understood too well that the capacities of an international organization often did not amount to more than the sum of its national members. Without the affirmation by national governments, without their consent, the League did not dare touch on the sensitive question of calendar reform. In 1927, it relegated calendar reform back to the national level by urging member states to form national bodies, committees of inquiry, for the study of calendar reform.[32] The League's Communications and Transit Section held its fourth international meeting in 1931 and finally organized a conference on calendar reform as part of that event. League bureaucracy was slow, and with so much emphasis on consensus and multilateralism, decisions took caution and time. The timing was unfortunate, however, for the conference of 1931 concluded that against the backdrop of worldwide depression, calendar reform had to take a backseat in favor of more pressing concerns.

During the 1930s, the League of Nations entered a new phase of its existence. The Nazis came to power in Germany and left the League in 1933, as did Japan; Italy under Mussolini followed in 1937. The optimism of the mid-1920s was rapidly fading, and with it, the hopes that internationalism and the League could prevent another European war died a slow death.[33] The League maintained its subcommittee on calendar reform and continued to communicate with national committees while collecting documentation on calendars until the late 1930s, when in the face of tepid feedback from members and growing international tensions in Asia and Europe it temporarily removed calendar reform from its agenda.[34]

• • • •

IN THE INTERWAR YEARS, those arguing in favor of a reformed calendar deployed a variety of rhetorical strategies. "Order" and "harmony" were frequently uttered in connection with calendar reform, as was "stability," a buzzword of the 1920s and 1930s in many regards.[35] "Rational" was an adjective that many self-styled reformers preferred for characterizing their plans. Calendar reformers also praised the "simplicity" of a rearranged calendar in an increasingly complex world.[36] In other cases, the now seasoned trope that the world was becoming smaller continued to serve as an argument. A Japanese ambassador to the United States found that "one of the characteristics of advancing civilization" had been "to increase man's power of moving rapidly from one place to another, thus minimizing his sense of space." In such a shrinking world, one calendar for all was the only sensible choice.[37]

Most prevalent among League officials was the notion that a common calendar was an instrument of world peace. Others shared this sentiment. In a telegram to the League's calendar conference in 1931, Mahatma Gandhi announced his support for a unified calendar, "as I am in favor of a uniform coinage for all countries and a supplementary language—like Esperanto, for example—for all peoples," he explained. National jealousies and myopia should not hamper reform activities, he demanded. One of the main German spokespersons for calendar reform designated one day in the new calendar "World Peace Day" as a reminder that all men were brothers. The language of "one calendar for the world" grew particularly loud once the guns of war threatened to sound again in Europe in 1938 and 1939.[38]

World peace was easily drowned in another chorus of voices, this one touting the economic advantages of a reformed calendar. From the early twentieth century to the 1920s and 1930s, a slew of arguments about the value of comparable numbers, formerly confined to a limited audience of accountants and statisticians, slowly entered the political mainstream and with it, public opinion. Cost accounting had become an important factor in the corporate world with the growth of complex multidivisional firms in the second half of the nineteenth century.[39] For the purpose of accounting, it was critical to use quarters with equal numbers of business days, units that could be compared without undertaking a lengthy reorganization of data. Business interests, organized in chambers of commerce and manufacturing associations or independently, now urged calendar reformers to tackle the unwieldy subdivisions of the current calendar arrangement.

Among those lauding the advantages of a uniform calendar for the business world was George Eastman of Kodak. In the late nineteenth century

Eastman invented a method for making photographic negatives by using gelatin emulsion on film instead of glass plates and developed Kodak, a mass-produced, easy-to-operate camera for the amateur photographer. For sales and distribution, Kodak, headquartered in Rochester, NY, created a worldwide marketing network of branch offices while setting up production and servicing facilities in Britain.[40] Perhaps due to his immersion in a number of different national business environments, Eastman grew increasingly interested in calendar reform. A representative for Kodak at the meeting of the National Industrial Council (a manufacturer's association) in 1928 reiterated Eastman's own views: "'Modern business management . . . which scrutinizes every detail of operations with a magnifying glass, needs every aid in production, sales, economics in overhead and so on that can be had. But the basis of all its operations, the element of time, is so unscientifically measured by the present calendar that it is a handicap instead of an aid.'" The representative held forth that "business management tries to build its economic structure by exact designs but it has to do it on a continually shifting foundation."[41]

Herbert Parker Willis, a professor of banking at Columbia University, similarly wrote in an article in 1931, "Statistics has of late years grown tremendously in its useful service to business, and today few industries are without their statistical measures." Eastman himself spoke of a "new economic age" that "must control its activities and to do so it must measure the performance of the past and future in units of time."[42] The International Association of Railways through its committee for accounting and currencies similarly pointed out that "the main advantages of a more uniform division of the year into segments of equal length (months, quarters) accrue for statistics. Without truly reliable and comparable statistics, effective monitoring, leadership, and control of a business is today unthinkable."[43] The International Chamber of Commerce's delegate to the League of Nations held that "the experience of the modern world seemed to show that the development of industry and commerce called for a system, rendering it easier to compare figures from week to week and from month to month." Comparison ranged paramount. Economic arguments for calendar reform made ample use of visualized data in the form of graphs, a new but now increasingly common feature among economic experts, in economics textbooks, and certain parts of the government in these years. Dramatically spiking and seesawing lines zigzagging along the temporal axis captured the numerical variation in workdays among different months, the abhorred "irregularity" and incomparability of the calendar, even more impressively than just words.[44]

The truth was that many corporations in the United States already observed a separate economic calendar. By the 1930s, a calendar with thirteen months of equal length was used for accounting purposes in a variety of prominent businesses, among them Eastman's own Kodak and Sears, Roebuck and Company.[45] Big business and banking were prominently represented among the numerous members of the American Committee on Calendar Reform to ensure that such a practice would become even more widespread. Members included Gerard Swope (as president of General Electric); Alfred P. Sloan (president, General Motors); Henry Ford (president, Ford Motor Company); and George Roberts (vice president of National City Bank of New York). The committee was large and besides corporate and banking interests included representatives from the government (Treasury Department, Bureau of Standards, Naval Observatory, Department of Agriculture, Department of Labor) as well as other influential but casually selected figures such as Adolph S. Ochs, publisher of the *New York Times,* as well as the dean of Harvard Law School and the president of MIT.[46]

After conducting a large survey of calendar problems in the late 1920s, the American committee established that over 98 percent of returned questionnaires opted for the thirteen-month plan, enough to convince the American calendar reform committee to officially endorse this format over others.[47] The choice was not entirely surprising, for the thirteen-month option was backed by George Eastman himself; Eastman was bankrolling the American committee. More surprising perhaps, among business groups there was a conviction that citizens would have no qualms with the major amendment of adding a thirteenth month and would get accustomed to the significant change of habit involved. To organized capitalism, the prospect of rendering comparable what to date was incommensurable overrode feasibility concerns.

Without government support, capitalism could only achieve so much in the way of reform. It was only once the political interest in calendar time peaked that business groups suddenly found an open ear for their calendar woes. The national calendar reform committees that had been set up at the request of the League in many countries enjoyed close ties with governments and, above all, economic experts in the service of governments. National governments, too, discovered the value of comparative numbers during these years. Extensive government statistics began with nineteenth-century census data and demographics but eventually comprised much more than that—numbers on crime, alcoholism, and other "social diseases" and, ultimately, economic statistics. But it was a long path from these numbers to the data wielded by governments today in the form of GDP or the consumer price index. Statistics itself had to emerge as an academic field.[48] Not coincidentally, one of the earliest internationalist movements of the nineteenth century was that of the community of statisticians. Their aim, among other things, was to establish criteria that would make statistics comparable across different national traditions. An international statistical congress was held as early as 1853, and in 1885, the International Statistical Institute was set up in The Hague. Statistics became the meta-science of comparison.[49] During and after World War I, national governments discovered the importance of data on nutrition; more generally, they used data for economic forecasting and planning and, on the international level, for the question of German reparations and war debts.[50] The National Bureau of Economic Research in the United States, the British Institute of Economic and Social Research, and the German Institute for Business Cycle Research (later renamed as the German Institute for Economic Research) were all founded during the interwar years. The League of Nations' own Economic and Financial Section was conducting massive efforts at data gathering and processing. Fortunately for the business proponents of calendar reform, then, governments' newfound preoccupation with comparable national economic numbers gave a boost and an air of necessity to their efforts. Calendar reform much more than the adoption of uniform mean times was widely framed as an economic matter, as the various national committees formed in the United States and elsewhere bespeak. The German committee for calendar reform, set up in June 1930, was situated at the German Institute for Business Cycle Research, whose director was simultaneously the head of the Imperial Office of Statistics.[51]

• • • •

THE LEAGUE, business groups, and national calendar reform committees were certainly among the most visible spokespersons for calendar reform, but they were not alone. Early on, internationalism had offered women a field of political activity; some of the most prolific nineteenth-century international NGOs were women's associations campaigning for suffrage and temperance alike. Several decades on, the Geneva "scene," as it emerged in the 1920s with its several NGOs now residing in the city on the lake, allowed women loosely affiliated with these organizations to pursue diplomacy. The League itself counted women among its personnel, if not in leadership at least in secondary positions. Essy Key Rasmussen, daughter of a well-known Swedish feminist, functioned as one of the League's liaisons with affiliated NGOs and calendar reformers while employed at the League's Secretariat for several years.[52]

Many of the countless calendar reform societies formed in Europe and America remained relatively obscure and never acquired notoriety beyond a limited group of calendar aficionados. One exception was the World

Calendar Association (WCA) founded by Elisabeth Achelis in 1930. Born in Brooklyn Heights in 1880, she was the daughter of the former president of the American Hard Rubber Association, and upon her parents' death inherited a fortune. Achelis never attended college and spent her youth as a girl from a wealthy background enjoying dances, parties, card games, concerts, and piano lessons, as she characterized her upbringing later in life. During World War I, she lent her time (and her car) to the American Red Cross Home Division, but other than that Achelis had no experience with philanthropy or volunteerism. In the summer of 1929, Achelis was mingling with other socialites at Lake Placid Club, where she attended a lecture by Melvil Dewey, inventor of the eponymous library classification system, on the topic of calendar reform; Dewey was singing the praises of the thirteen-month plan.[53]

According to her own stylized autobiographical narrative, this was a transformative experience: Achelis decided to devote herself to a cause, and that cause was calendar reform. Over the following two decades, Achelis spent her time and money promoting her "World Calendar," a variation on the twelve-month scheme that would make quarters more even by rearranging the number of days in months and intercalating a blank day. The World Calendar Association consisted primarily of herself, an office assistant, and later, a director who took over the main duties of correspondence and bookkeeping. The main task the association set itself was to relentlessly correspond with myriad businesses, governments, professional associations, and reform societies and urge them to espouse her World Calendar proposal. By June 1931, Achelis stated her association had approximately 2,500 members the world over. Besides voluminous promotional material prepared by the WCA, Achelis authored a book and, between 1930 and 1955, published the bimonthly *Journal of Calendar Reform*.[54]

During the 1930s when the League's Communications and Transit Section was holding its calendar conference and when calendar activism was at its height, Elisabeth Achelis spent months in Geneva, residing at the Beau Rivage, hovering in front of office doors in the corridors of the League of Nations Palace, and enjoying dinners with diplomats and statesmen. From Geneva, she undertook several trips to Europe, where "The Calendar Lady," as she was now known, met with diplomats and members of the various national calendar reform committees. More adventurous and defying her distaste for the conditions of early air travel, she swapped London, Paris, and Rome for long trips to Asia and the Middle East.[55]

Achelis's male counterpart and occasional nemesis was the British engineer Moses B. Cotsworth. The other ardent reformer who would head a calendar NGO previously worked as a statistician for the London and

North Eastern Railways in Britain, where he designed a system by which the cost of transportation was analyzed for compiling railway rates and charges. His work frequently required Cotsworth to ascertain the difference between earnings of preceding months or the equivalent month of a past year. It was in this context that he discovered the advantages of a thirteen-month calendar with identical twenty-eight-day months for business statistics. He compiled his ideas into a book as early as 1902, but it was only in the 1920s that Cotsworth quit his job to become a full-time calendar reformer. He founded the International Fixed Calendar League to promote his ideas and soon found a solvent sponsor in George Eastman.[56]

The League of Nations relied on self-supported individuals like Cotsworth and Achelis whenever it was too thinly spread or simply recognized an opportunity that came free of cost. Cotsworth coordinated his foreign trips with the head of the Communications and Transit Section as the chief responsible official for calendar reform and updated the League personnel regularly on his whereabouts.[57] In the fall of 1930, Cotsworth wrote from Tokyo, where he met with a German diplomat to discuss the progress of the German national calendar reform committee. Stops in Singapore, Calcutta, Bombay, and Cairo followed on the same trip. A journey through Central and Southeast Europe took him to Vienna, Prague, Budapest, Belgrade, Bucharest, Sofia, and Istanbul. The relationship between the League and its volunteers was not without tensions, as NGOs only loosely associated with the League of Nations often used the League's name in advertising their cause, even acting in its name. Since the League's calendar diplomacy was always a delicate tightrope act between the churches and different national interests, too much activist entrepreneurship was frowned upon in Geneva. But as long as Cotsworth and Achelis limited themselves to reconnaissance missions and taking the temperature of public opinion abroad, they were most welcome in Geneva.[58]

Reform plans from individuals like Cotsworth and Achelis were joined by a large number of submissions from individuals and smaller associations. Between 1924 and 1931, roughly 560 reform projects were received in Geneva.[59] After League officials sifted through all those materials, eliminated duplicates, and classified submissions into broad groups, the League's calendar reform committee had before it three piles. One contained material that merely considered an equalization of the year's quarters through shifting days from one month to the next. In a second pile there were those calendars that went a step further and established perpetuity by reducing the number of ordinary days in the year from 365 to 364, dividable into four quarters of ninety-one days each. To maintain the astronomic length of the year, these schemes added a supplementary day outside the order of the week. The third pile was reserved for calendars that added a thirteenth month, though chances for their adoption were low from the start—the League committee surmised early on that "the division of the year into thirteen months would be rejected a priori"; since suspicion of the number thirteen was pronounced, this assessment was probably right.[60]

• • • •

MEMBERS of the League's bureaucracy time and again emphasized their neutrality and declared not to be speaking for their respective countries of origin. But as soon as national governments seriously contemplated the study and implementation of calendar reform, the universal calendar was suddenly rebranded as a national effort. Calendar reform as advanced by the League of Nations depended on national governments for publicity and legitimacy at home; internationalism had no concrete meaning unless it was translated into the requirements of domestic politics.

In the United States, a possible reform of the calendar was fused with the politics of the New Deal. Prior to that, Congress had twice dealt with calendar reform but never pursued such plans more vigorously. In 1922, a bill was brought before Congress proposing the adoption of a thirteen-month calendar, but hearings on the subject did not result in any action. The arguments advanced in 1922 centered more on standardization in general than on economic advantages and statistics. One of the self-declared calendar experts testifying before the committee, the head at the Liberty Calendar Association in Minnesota, stated that the greatest benefit of reform would be to create a standard month. "We have no standard month," he went on, "even in this day of standardization." Moses Cotsworth also gave testimony before the commission, arguing similarly that "almost everything for our life at the present time is standardized, yet we do not have a standardized month. Our times for earning and spending are unequal." Economic reasoning was not yet dominant.[61]

Calendar reform as standardization was moreover discussed as a Pan-American project. In 1915, a Pan-American financial congress had commended the standardization of commercial laws and regulations currently in use in the Americas. Such policy recommendations resulted in the First Pan-American Standardization Conference held in Lima in 1924. Standardization had become so popular in the United States that it was now held to occupy an important "place" in "American life," as a publication announced.[62] Prompted by the Lima conference, the Pan-American Conference in Havana in 1928 passed a resolution that called on member countries in the southern hemisphere to instate national committees on calendar reform. Spurred by the events in Havana, yet another bill therefore came before the House of Representatives in 1928, this one calling on the American president to hold an international conference on calendars, but it failed to generate support.[63]

It was not until the Great Depression hit and the New Deal was born that government officials suddenly found compelling reasons to consider calendar reform more actively. When the National Recovery Administration (NRA) was established, it was tasked among other things with collecting a wide range of industrial data and with reclassifying existing numbers submitted in the past but found to be incomparable. The idea was to create a common basis of data reporting, and one element of such a shared foundation was the possibility of utilizing uniform time spans for reporting and processing data. In the fall of 1933, M. B. Folsom, chairman of the Committee on Calendar Reform of the American Statistical Association, in a speech on "the use of four-week reporting periods for industry with relations to the recovery program," recommended that four-week periods should be introduced for reporting industrial and trade information. Folsom explained that weekly, biweekly, and four-weekly periods were already widely used as the basis for payrolls, and this informal "calendar" arrangement ought to be applied to other areas as well.[64]

The speech and the findings of the American Statistical Association were forwarded to the Central Statistical Board, a New Deal Agency. In October 1933, the Board arranged a meeting of several organizations and agencies to discuss calendar reform. The resolution concluding the meeting opined that "much confusion and misinterpretation arise from the lack of uniformity in the time units for which data are given," a condition that complicated the "use of series, collected at frequent intervals, of statistics of employment, payrolls, production and other data used to measure the volume of business and to determine economic trends." As the resolution explicated, "such statistics of business, when presented on the basis of calendar months, lose part of their value and are liable to misinterpretation, from the inequality in their number of working days in the several months, and from the fact that the month is not a multiple of the week." This was particularly regrettable for "at the present time in connection with recovery programs there is taking place a marked expansion in the scope of current statistics of business." Now was therefore a "favorable occasion for improving practice with respect to the time-unit basis of such statistics." The Board therefore recommended the introduction of four-week periods for reporting industrial and trade information. Throughout the 1930s, calendar reform and its advantages for statistics and economic recovery was a recurring talking point in the press and among authorities alike.[65]

• • • •

ANOTHER INSTANCE of nationalizing universal calendar time took place in postindependence India. As long as the subcontinent was in large part ruled by the British, calendar reform did not find much of an audience in India. When the League of Nations was investigating public opinion on calendar reform in the 1920s and 1930s, the British administration cautioned against attempting to meddle with calendars. The majority of Indians followed various astronomical calendars and would refuse a reform as radical as any of those proposed. Possible benefits of the reform would thus be offset by the indifference with which local populations were sure to greet a reformed calendar. British officials, perhaps wary of tampering with time after the experiences with Greenwich time, shied away from the question of calendars.[66]

Once India gained independence in 1947, attitudes toward calendar reform changed. Unifying the several calendars in use on the subcontinent now turned into an aspect of nation-building and modernization as pursued by India's new leadership under Jawaharlal Nehru. In 1952, the Nehru government appointed a committee for the study of calendars, situated at the India Council of Scientific and Industrial Research. The committee was charged not only with studying India's numerous calendars but also with establishing an "Indian Greenwich," an observatory located on the meridian of the existing "Indian Standard Time" of UTC + 5:30, in charge of unifying clock times in India. Currently, the Hindu civil day was counted from sunrise, whereas Muslims considered the day to begin at sunset. All over India, certain calendars moreover reckoned in lunar days computed from moonset to moonset. The new national observatory was supposed to bring these diverging practices into the fold. "The need for evolving some uniformity is clear as crystal," as Keshev Deva Malviya, India's first petroleum minister who was part of the effort, proclaimed. The government faced problems such as determining the days of public holidays, for the same religious festival was observed on different dates in different localities.[67]

Obtaining calendar unity easily blended with the drive by the modernizing state to extinguish the "countless myths, religious dogmas, superstitious and astrological practices [that] have damaged the cultural unity and development of the country, through their connection with calendar observance." The committee published its first report in 1955, with Nehru himself authoring the introductory comments to the committee's findings. The Indian leader now set the number of estimated calendars at thirty, a condition he viewed as "the natural result of our past political and cultural history and partly represent[ing] past political division in the country." Nehru stated, "Now that we have gained independence, it is obviously desirable that there should be a certain uniformity in the calendar for our civil and social purposes."[68]

• • • •

IT WAS GERMANY'S national reinterpretation of calendar reform that most dramatically revealed the limits of internationalism and universalism. When the League of Nations began its examinations, German observers followed with great interest. The German national calendar committee began its work of probing public opinion as suggested by the League after 1926. With the Nazis' successful bid for power in January 1933, calendar reform remained popular, though under entirely different auspices.

Since the late nineteenth century, the German Imperial Office of Statistics published a so-called Basic Calendar. The booklet contained the annual list of Protestant and Catholic church holidays and the calendar for the ecclesiastical year as well as lists of name and saint's days. Private publishers then used the official data to compile their own calendars. In the summer of 1933, some of these publishers turned to the Nazi Chamber of Literature and the Imperial Office of Statistics to express their dismay at the presence of "Jew names" in German calendars. Name lists in the German Basic Calendar contained names such as Abel and Seth; even common names such as Josef, Maria, Susanna, Elisabeth, Gabriel, and Johannes, they warned, were of "Hebraic" origin. In the official calendar of the Evangelical church, New Year's Day was moreover notated as the day of Christ's circumcision.

In the same summer of 1933, a few months after his rise to power, Hitler received correspondence from a German citizen declaring that "no time in world history has ever offered such a favorable opportunity for achieving a German act of culture as [the time] brought about by you," and calendar reform was such a deed, as "one of the most important definitions of time carries un-German names."[69] The head at the Nazi Chamber of Literature quickly declared it "desirable that uncommon Hebraic names would be removed from calendars and replaced with German names." A few months later, the Ministry of the Interior ordered for "Jewish holidays to no longer be listed in calendars used in government offices." Such inquiries by troubled publishers and concerned individuals set off an examination of calendars and name lists conducted by the Imperial Office of Statistics in cooperation with the Protestant and Catholic churches.[70]

Nazi calendar reform then turned into an endeavor to insert more German saint names into lists of names, resulting in additions such as "Adelhard,"

"Genoveva," and "Farhilde," Germanic-sounding names reminiscent of medieval German literature. To some, the initial purges carried out by the churches and Nazi authorities were insufficient. The Austrian publishing house Adolf Luser wrote to the Ministry of Ecclesiastic Affairs in May 1940, describing the inventory of name days as a means to "subtly guide mothers and fathers to give their children German names." Jewish names—that is, those that had been classified as such since 1938—"of course have no place in a German calendar." Some particularly "foreign" (in this case, foreign read Slavic) names such as "Bogislav, Ladislaus, and Kasimir" ought to be removed too.[71]

Advocates of Nazi calendar reform were not prepared to stop with lists of names and name days but eventually wished to see month names changed as well. In keeping with the Nazis' attempt to invent a Nordic-Germanic past and adopt elements of that history in Nazi symbols such as the emblem of the SS (based on runes from Germanic alphabets predating the adoption of Latin script), calendar months were now to carry designations such as "Hartung," "Hornung," "Lenzmond," and "Ostermond," Germanic names suggesting a close connection between the seasons, the state of the soil, and agricultural rhythms as in "the month of the spring moon" and "the month of severe frost."[72] Such proposals did not come to pass. But the League of Nations' internationalist efforts at furthering the adoption of a peace-promoting, universal world calendar had morphed into moves to purge the calendar of "Jew names" and make it more "Germanic."

• • • •

FROM THE ONSET, religious authorities of different hue had been skeptical if not outright hostile to the notion of calendar reform. It was their steadfast opposition that made a change in calendars an increasingly unrealistic endeavor. In 1931, the League of Nations conducted a survey of religious opinions. Broadly speaking, Protestant churches were more willing to discuss the stabilization of Easter than Catholic authorities, yet the reshuffling of days through the addition of a blank day, and certainly the creation of a thirteenth month, were off-limits. The Eastern churches normally made their support dependent on a collective embrace of a reformed calendar by all major churches.[73] Silently, everybody from the Protestant churches to national governments to the League of Nations acknowledged that the real linchpin was the Vatican. Calendar reform was understood to be impossible without Rome's assent, a testimony to the sway that Catholicism was believed to hold over minds and hearts even in a presumably more secular age. In 1931, Vatican officials told the League that the Easter question fell entirely within its own purview and was ecclesiastical in nature, and that since the League was obviously out of line in even addressing the question of calendars, the Vatican would abstain from sending observers to Geneva for conferences or committee work.[74]

Religious groups of varying denomination were conjoined in their objection to calendar reform due to the intercalation of a blank day outside the rhythm of the week. The various thirteen-month plans were never considered a serious option, for such an intervention in the calendar was deemed too radical, too deviant from religious scriptures and traditions. But even the more limited amendments proposed by twelve-month plans overturned what some viewed as inalterable tenets of their faith. Intercalating a day without a weekday name as a "blank day" or "world holiday" meant changing the sequence of days. Hence what used to be Saturday or Sunday would no longer fall on the "true" Saturday or Sunday but would "migrate" through the week. In essence, this was a variation on an objection that had been made against mean time and summer time for so many decades. A Sunday would not simply move to a different day while the "old" Sunday obtained a new name. Instead, the "true" Sunday would fall on what was now called Monday and so on, just as human rhythms of sleep and nutrition intake would not simply move in accordance with the time shift but stay put. As the Religious Liberty Association (affiliated with the Seventh-Day Adventists) explained in a pamphlet titled "Calendar Change Threatens Religion," under a "highly financed scheme which would abolish religious days," an "entirely different day would replace Sunday should our calendar changers succeed in their designs. It would not be Sunday at all. It would actually be Monday. But it would be given Sunday's name. And those who now hold Sunday as a holy day would be asked to accept Monday in its place, re-christen it Sunday and observe it instead of the real Sunday."[75] Since Muslims, Jews, and Christians shared an adherence to one particular day of the week as a day of rest or worship, the three religions of the book were fairly united in their rejection of calendar reform. Christians and Jews often referred to the status of Friday in Islam when stating the centrality of Sunday or Saturday to their own faith.[76]

Among the most outspoken critics of calendar reform was Joseph Hertz, chief rabbi of the United Hebrew Congregations of the British Empire. Time and again, Hertz stressed the vital interest of Jews in the regular sequence of the seven-day week. The British rabbi published several pamphlets and books against calendar reform, including one titled "The Battle for the Sabbath at Geneva," calling on every Jewish community to participate in a "Holy War" against the abomination that was calendar reform. Hertz moreover explicitly attacked the economic and financial bent of pro-reform arguments. A reformed calendar carried the imprints of "American financial interests," Hertz found, and was being promoted by "the power of American propaganda which stretches all over the world, shies no expenses and works scrupulously." It was an attempt to "Americanize and mechanize life," as Hertz put it, perhaps alluding to the global allure of Fordism in the interwar years.[77]

Hertz was joined by other Jewish groups. The Central Conference of American Rabbis corresponded regularly with British and European Jewish authorities over the matter. The Jewish Sabbath Alliance of America, attempting to convince unions and factories to observe five-day weeks to facilitate Sabbath observance, spoke of the "disastrous consequences" for observant Jews in case the rhythm of the week was interrupted. A Jewish group from Vienna bemoaned that, whereas such units of time as seconds, minutes, hours, days, months, even years were declared results of science or custom, the seven-day week was taken to be "purely religious" by Orthodox Jews; worldly authorities had no solid grounds for meddling. Contrary to the Holy See, Jewish groups actively sought accreditation at the League's conferences and lesser meetings. In 1930, the World Union for the Safeguarding of the Sabbath held its own congress in Berlin and discussed calendar reform in preparation for the upcoming League event in 1931. Joseph Hertz and others had established the "Jewish Committee on Calendar Reform" that was slated to make a statement before conference attendants. Discussions made it unmistakably clear that "Jews are unalterably opposed to any change in the reckoning of time" that would interrupt the continuity of the seven-day week and thus send the Sabbath, a "fundamental institution in Judaism," coasting through the week. Such a "wandering," "nomadic Sabbath" was unacceptable.[78]

Correspondence about calendar reform mostly originated in Europe and North America. But the occasional non-Western viewpoint found its way to Geneva, a testimony to far-flung channels of communication and intellectual exchange. The "Association of Muslim Scholars" from Rangoon, Burma, submitted a protest letter to Geneva in which it explained the strong objections by Burmese Muslims to any alteration of the weekly cycle from a religious and Islamic point of view.[79] G. A. Chowdhury, writing from a literary society in Delhi, India, explained that Friday was a day of special obligations to Muslims and that any blank day would make the Friday of the new calendar unacceptable for Muslims as "their" Friday. Changing Friday as the modified calendar stipulated was not permissible from the religious point of view, and Muslims would have to keep a separate calendar, thus in turn making the whole point of reform—simplification—obsolete. Chowdhury predicted widespread protests from

Muslims as well as Christians and Jews should such a reform come to pass.[80]

• • • •

AFTER CALENDAR ACTIVISM reached its height in the early and mid-1930s, it faded. Following World War II, two less influential countries (first Peru, then Panama) made attempts to place the calendar on the United Nations agenda again. In 1953, India made an advance and requested a discussion on calendar reform during the upcoming session of the Economic and Social Council (ECOSOC), clearly motivated by its own internal efforts to homogenize the country's multitude of calendars. When the matter at long last found its way onto ECOSOC's agenda in 1955, it was suggested to postpone a debate as responses from governments had been tepid, and the United States in particular had expressed strong opposition, citing religious concerns. In 1956, the League concluded that universal acceptance of any such reform was highly unlikely due to the undesirable effects on many aspects of religious life. The representatives at the Council meeting voted to adjourn further discussion of the matter sine die.[81]

Calendar reform never had more than a limited appeal due to the underlying Gregorian tradition of the proposals for amendment. Non-Western societies certainly witnessed a mounting interest in calendars briefly after the turn of the century, partly in the Gregorian, partly in indigenous calendars, partly even in a reformed "World Calendar," if only to oppose it. Muslim scholars discussed the timing of Ramadan. Hindus and Parsis in India were considering a reform of their calendars.[82] A man stating his name as Joseph Pohjey Hsue, writing to the League of Nations from the French concession in Tianjin, China, devised an eternal calendar that was more or less identical in structure with Cotsworth's thirteen-months plan. The manager of a Shanghai newspaper mailed a hundred copies of a Chinese leaflet on calendar reform to Cotsworth for distribution. In China, the Confucian Association of Beijing in 1918 brought a proposal before the Chinese Senate on the establishment of a global government "for the harmony of all nations of the world." This new world order called for a universal calendar that would mark a new era beginning with the inauguration of the new government.[83]

Euro-American calendar reformers barely acknowledged these alternative conversations about changing the calendar. The fact that countries such as Japan, China, and after 1917, Russia had already adopted the Gregorian calendar was usually adduced as confirmation for the assumption that non-Western societies would "voluntarily forsake their less useful calendars," as Moses B. Cotsworth stated. He elaborated that "about 380 millions of less cultured people have hitherto been handicapped by using crude local calendars beginning at different dates in the solar year, and are now realizing the practical advantages obtainable by use of the more fixed Gregorian calendar." Cotsworth and likeminded others were wrong, as the default solution in non-Western parts of the world seems to have been (and still is) a use of local calendars alongside the Gregorian calendar.[84]

In the end, however, what rendered calendar reform unsuccessful was not a lack of appeal among those following calendars other than the Gregorian. The movement was brought to its knees not by non-Western resistance against the imposition of a "foreign" calendar time. Universal calendar reform failed right where it originated, in Europe and North America. Religious authorities successfully alerted public opinion to the religious objections to the project; their protests made national governments reluctant to interfere with religious sensitivities encapsulated in calendars. With the help of "modern" communication and internationalist organization, organized religions ultimately succeeded in generating enough publicity to scare off governments. Religious authorities beat the League and calendar NGOs at their own game.

Religion itself had been transformed in nineteenth-century globalization. Under pressure from nation-states, empires, and colonialism seeking to stymie the influence of religion, different denominations lived through a striking process of reinvigoration and reorganization. This revival was facilitated not least by the forces that shaped an interconnected world in general: the proliferation of affordable print products allowed scattered diasporas to imagine themselves as conjoined with fellow religionists in faraway places. Railways and steamships dispatched a growing number of missionaries overseas while offering pilgrims of different religions the opportunity to travel faster and cheaper. The globalization of religions heightened a sense of a worldwide community through events such as the World Missionary Conferences in Protestantism, the papal Jubilees of Roman Catholicism, and the World Jewish Congress.

The dominant world religions moreover established international organizations aimed at fostering more self-conscious global identities among their members. Often, this involved a turn to social activism on an international scale.[85] The ecumenical movement took off at the end of the nineteenth century, with the World Parliament of Religions convening in Chicago in 1893 and the World Missionary Conference in Edinburgh in 1910. Several predecessor organizations of the World Council of Churches (founded in 1948) were established after World War I. One of them, the Universal Christian Council for Life and Work, took its seat in Geneva, the site of several internationalist Christian meetings of the period. Many religious internationalist groups were affiliated with the League of Nations, where they partook of the internationalist scene of the interwar years. Religious officials effectively combated calendar internationalism with its own weapons. Public opinion and internationalist connections were paramount in these efforts, as the relentless activism by a globally operating Religious Liberty Association (RLA) illustrate. At home in the United States, the RLA bombarded governments and citizens with countless pamphlets, advertisements, and letters intended to raise awareness. The "international arm" of the Seventh-Day Adventist Church similarly opposed calendar reform as restricting religious practice. In 1931, the RLA wrote to the League of Nations from its Indian headquarters in Poona, submitting scores of pages covered in signatures collected on the subcontinent against the adoption of a world calendar. Dotted down in Latin, Perso-Arabic, and Indian alphabets, these petitions—the Religious Liberty Association claimed there were more than 14,000 of them—demanded the Christian, Muslim, and Jewish right to observe the "unbroken succession of . . . fixed religious days."[86]

The failure of calendar reform was furthermore precipitated by dwindling European-internationalist spirits. Internationalism paired with a global mentality had generated much of the enthusiasm for a reformed calendar in the decade before the war and in the 1920s and early 1930s. It evaporated when the rise of fascism and Nazism and a war of unprecedented scope and destruction ravaged the world between 1939 and 1945. After World War II, the Cold War soon appeared on the horizon. To paraphrase the Austrian novelist Stefan Zweig's 1942 farewell to the nineteenth century, internationalism and self-conscious globality were "yesterday's world."[87]

Economic reasoning and capitalist interests alone were unable to carry over calendar reform into a new era where the global mind-set of the nineteenth century no longer inspired people. When calendar reform petered out in the late 1930s and again in the 1950s, the interest of both business groups and governments in comparable, homogenous units of calendar time had by no means disappeared. The need for comparable statistics and generally for commensurable quantifications of what was now referred to as "the economy" arguably increased after the war.[88] National income accounting and the very idea that something like the entire economic activity of a country could be measured was by now a fairly common operation in many countries. An interest in comparable numbers that grew out of a New Deal effort to ascertain economic performances and the cost of living in times of crisis subsequently served war-related planning during the conflict and the reconstruction of European economies thereafter. In Europe, already while the war still lasted, economic and social data on growth and what was coming to be known as "standards of living" stood central in conceptions of a welfare state. At the newly founded United Nations, the systematic compilation of comparable data from what were now dubbed "developing" countries became a top priority. Yet despite the persisting, even growing utility of improved monthly and quarterly accounts, rearranging the calendar was no longer a serious option. In a way, capitalism had adapted in a characteristic show of flexibility and with the adaptability that had facilitated capitalism's spread around the globe in past centuries: many businesses and even governments today use fiscal years that differ from the calendar year for accounting purposes. The use of multiple, parallel orders of coexisting times proved a solution that offered some of the benefits of a rearranged calendar to economic interests while at the same time avoiding social conflict over cherished religious and cultural markers of time. The economic quality of homogenous time was more central to calendar reform than it had been to the reform of clock times, yet in both cases, the flexibility and variety of capitalist forms of economic organization should not be underestimated.

The all but disappearance of calendar reform underscores how utility and practical need never played more than a secondary role in the ideology of universal time unification. The much-lamented absence of uniform clock times and the stated need for more accuracy and regularity in an interconnected age bore little foundation in reality in a world in which time distribution remained unreliable and partial as late as the 1930s and 1940s. Interconnectedness did not necessarily require homogeneity to thrive. Inversely, even though comparable numbers were more important than ever before in a variety of public and private contexts after World War II, calendar reform had become unsustainable, for something else was amiss. The Cold War world lost trust in the cooperative tune of nineteenth-century internationalism, however limited and Euro-American it had been. Instead, those with the power to do so increasingly took matters into their own hands. The United States in particular was unwilling to entrust the "international community" with issues it considered of utmost importance. The favored solution was now to shape internationalism in such a way that American interests would be safeguarded, to govern the world "the American way."[89] Yet the intellectual ingredients of nineteenth-century internationalism and time reform—a universalist creed, an almost utopian belief in universal solutions for global improvement, a technocratic vision of a malleable world that could be engineered into shape, a notion of universal time and progress against which the achievements and shortcomings of different societies were measured—lived on in the countless "modernization" projects that Western social scientists and their executors now bestowed on the "developing world" with its benighted peoples who, disadvantaged by history, race, and geography, lacked the good fortune of inhabiting the same stage of historical time as the West.

# Conclusion

When global and transnational history proliferated in the graduate curricula of American universities, in specific journals, and in methodological calls for broadening the scope of historical inquiry in the early 2000s, this quest for a global and nonnational past was implicitly inspired by the concurrent realization that flows of data and goods through networks of technology, communication, finance, and trade had reached unprecedented scale and scope in the 1990s. The questions that historians ask of the past are, in some form or another, often products of their time.

Since the 1990s, the invocation of stable networks, flows, and presumably open information and data has taken on different incarnations, from talk about the so-called great moderation (the supposed stabilization of the business cycle from the mid-1980s) to, most recently, the uncritical hailing of the Internet, big data, and Silicon Valley–generated "smart" technologies as the cure to the world's most pressing problems. In many such debates, globalization rhetoric did and does not stop short at depicting a world of border-crossing, time- and space-defying flows. More importantly, flows and connections are interpreted in a tendentious reading as a one-dimensional process that reduces globalization to certain economic qualities. In the aftermath of the 2008 financial crisis, *The Economist* for one frequently warned against scaling back "globalization," and by that meant restricting the flow of capital through protectionist measures.[1] Such an interpretation not only has clear definitional limits in espousing a notion of globalization as merely economic but also points to the fact that "connectivity talk" is always more than just an analytical assessment and neutral description of the present. Here, the evocation of flows and connections serves to promote a particular brand of free-market capitalism.

The history of globalization as explored through the lens of time reform in the second half of the nineteenth century lays bare the limits of such a purportedly descriptive narrative of connectivity in various ways. When nineteenth-century onlookers began picturing the world as a global village in which uniform clock and calendar time would render differences comparable, they talked and wrote about the urgent need for a system that remained incomplete for decades or, in the case of calendars, never came to pass. Transportation and communication may have run more smoothly in a world with twenty-four uniform time zones, but in the absence of uniform time they worked just fine. So did capitalism, which continued to thrive on informal temporal arrangements and heterogeneous clock and calendar times. If Sven Beckert's masterful global history of cotton offers any indication, flexibility might in fact be capitalism's greatest asset.[2]

"Connectivity talk"—the evocation of networks, connections, and instantaneous communication as a characteristic of both today's world and that of the late nineteenth century—is better understood as an ideological formation. An uncritical global history adopts some of these tropes too readily by focusing exclusively on the spread and flow of ideas and practices, and on border-crossing, continent-straddling movements. Such a version of global history risks making itself the mouthpiece of an ideology that portends to merely describe a networked globe but more often seeks to remake the world in its highly normative mold. Today's apostles of interconnectedness are heirs to nineteenth-century visions of hegemony. The invocation of networks and connections today resembles the proliferation of "time talk" among Europeans and Americans around 1900, which was just as politicized and nonuniversal, nonneutral. When Europeans and Americans wrote about time- and space-defying connections and uniform time as a means to bring order to a globalizing world, they proposed to create a world in their own image and a world of their own domination. That world was by no means "flat" but strikingly hierarchical. It saw the rise of the core regions of industrialization at the expense of the periphery and the expansion of imperial and colonial rule, developments that entrenched economic and political differences and unequal disseminations of wealth and power. The vision of the nineteenth-century global village united in uniform time surfaced at a moment when, in many ways, the world was becoming not a more homogenous but a more uneven place.

In contrast with such narratives, the global history of time reform shows how uneven, slow, and full of unintended consequences interconnectedness was. Hence, it is conceivable (although admittedly not possible here) to write a history of globalization that focuses exclusively on those ideas that did not spread or move, on the limits to mobility, and on those places that even at the beginning of the twentieth century remained unconnected—peripheral to intellectual, financial, and trade networks. Such histories were equally and integrally part of globalization. Interconnectedness moreover shifted over the course of time. Localities fell in and out of globalization as political and economic geographies were rearranged. A historical perspective that views unevenness, slowness, and global schemes gone awry not as temporary impediments to the progression of globalization but rather as inbuilt structural features of that process has much to offer to observers of the dynamics of contemporary globalization as well.

• • • •

THE HISTORY of nineteenth-century globalization followed no single pattern. In many realms, the world was becoming both more differentiated and more homogenous at the same time. From the bird's-eye view of history in particular, integration and fragmentation went hand in hand. The convergence of commodity prices and real wages among a few mostly North Atlantic world countries was accompanied by (and arguably based on) a more entrenched global division of labor as autonomous regional economies were drawn into the world economy.[3] As recent works on the global nineteenth century have shown, certain distinguishable types of cities, classes, or states emerged in the nineteenth century, similar in general nature throughout various parts of the world but different upon closer inspection. It lies in the nature of syntheses, with surveying character, that process is rarely part of all-encompassing global histories. Just how these global typologies of cities, classes, and states came into being in their various local and regional contexts is lost when viewed from above.[4] Moreover, local conditions, the importance of individual actors, and the subtle translation of contents disappear at the high altitude of experts and international organizations. Global and international history as well as longue-durée history easily become tip-of-the-iceberg history. Telling stories of entanglement and globalization therefore works best when global and international history consciously operate on different levels of scale, often ditching the global for specific national, imperial, and particular local urban archival contexts. It is from this angle that the uneven spread of interconnectedness and the unconventional vectors of transmission for ideas and practices appear most readily. What global historians writing about Europe as much as other parts of the world have to offer to the discipline of history is a combination of regional archival and language skills with a regard for "big structures, large processes, and huge comparisons."[5]

Yet even careful analysis of scale and attention to unevenness and to the occurrence of occasional hybridizations of indigenous with distant thoughts and practices cannot detract from or mitigate European ascendancy in the nineteenth-century world. Nineteenth-century globalization and, in extension, the globalization of time were predicated on European dominance and, in turn, the attraction of certain elements of European thought, science, and technology in other parts of the world. Non-Western societies were held in thrall to such supremacy. The global history of time reform is not one of resistance to Western temporal predominance, not even primarily one of hybridization (although hybridization did occur, especially in global cities such as Beirut). It is the story of how perceptions and apprehensions of an interconnected world gave rise to the politics of global time reform, and how time served as an intellectual and institutional device for imagining the world as global and interconnected. In this process, contrasting interpretations of the consequences and meanings of interconnectedness made for a much slower and much more uneven application of certain tenets of uniform time than commonly acknowledged. Unevenness persisted in the non-Western world as much as in Europe itself.

The apprehension of interconnectedness, a reflection on a world in which time and space dramatically changed shape, was not unique to Europe and North America. The way that world was depicted was. Europeans and Americans spoke of space- and time-defying communication and transportation, of a world in which traveling from Paris to New York was easier than traveling from Paris to Marseille had been a century ago. The French head of the geographical society of Oran in Algeria surmised in 1898 that, "In a way, the surface of the planet has shrunk. The habitat of humanity has become a small region that can be toured in a few weeks, and in which all parties instantaneously are put in contact with one another."[6] Under the same impression of interconnectedness, commentators in non-Western societies discerned a different underlying rationale: the world was indeed shrinking, for Europeans and Americans snatched up entire continents. Improved transportation and communication primarily placed the globe at European fingertips to do more of the same.

From the perspectives of onlookers in Beirut, Beijing, Cairo, Tokyo, and Bangkok, the late nineteenth- and early twentieth-century version of the global village was what remained after Europeans seized everything they managed to get their hands on. In the confines of collapsing global space, self-strengthening through the targeted adoption and translation of certain elements of Euro-American science, culture, and their apparent management of time was the only way to breathe freely.[7] These were engagements with an interconnected world on distinctly "national," "civilizational" terms. It has become commonplace to assert that globalization consists of interactions between the global and the local.[8] Local constellations certainly had a leading role in shaping the appropriation of time around the globe, whether in the Levant, British India, France, or Germany. But much more than commonly acknowledged, globalization accentuated national differences and drove nationally motivated engagements with time into the open. Over the course of some eighty years, the politics of adopting uniform mean times gradually added up to form a system of uniform time zones. But global time, despite its eventual success, was for almost a century national time. And due to religious and national concerns among governments and religious circles, calendar reform never came to pass altogether. Because the unification of clock times was more or less achieved around the mid-twentieth century after almost eighty years, the standardization of time is today understood to have been about clock times exclusively. In contrast, calendar reform is mostly forgotten because unlike the unification of clock times, it did not prevail. Yet in terms of the attention it garnered, the discussions it generated, and the interests it galvanized, calendar reform was in fact more prominent in its time than was the standardization of clock times slightly earlier, and at least as widely debated as daylight saving or summer time, if not more. It is important to remember that due to religious and nationalist or cultural pushback against calendar reform, time reform as a whole therefore remains incomplete to this day. Here, too, efforts to internationalize and globalize resulted in the insistence on national and religious difference.

In the case of Islam, the reconfiguration of global time and space wrought by the telegraph first brought the variations in determining the Islamic calendar into the light, as news about multiplying practices in the definition of the lunar months surfaced with regularity. In turn, technologies like telegraphy (and it has to be added, print) and their role in putting Muslims across several continents in contact led scholars of Islam to imagine the Muslim community not as a nation but as a global entity that drew strength from the uniformity of religious practice and the unison rhythms of calendar time. Yet the dissemination of certain ideas about how to determine the Islamic calendar also brought Muslim thinkers to naturalize and essentialize the imagined global community of Muslims as a holistic civilizational entity and to naturalize and essentialize this community in similar ways in which nations and races were essentialized and "invented" throughout the nineteenth century. In this, national and civilizational reorganizations of time resembled one another.[9]

Efforts to divert the connections of an integrating world to serve nations, nation-states, and civilizations were the norm rather than the exception. The reasons why uniform time zones and mean times advanced so slowly and remained so vastly incomplete until the middle decades of the twentieth century were varied, and the complications of imagining abstract time added difficulties. But one important factor was the nationalizing and regionalizing dynamic behind the adoption of mean times, in Europe and beyond. Decisions were made based on regional integration and, often, relations with neighboring countries. Yet the nationalization of time gradually led to the accumulation of more and more uniform nationwide mean times until by the mid-twentieth century, most time zones were set in even distances to the zero meridian at Greenwich. This was an additive and unintended process. The instrumentalization of efficient time management in the service of Arab and Islamic self-strengthening had similar unintended consequences. When societies in the Middle East and Asia strategically engaged certain facets of what they perceived as "Western" modernity, they did so on national or civilizational terms. But inadvertently, such strategies for self-improvement and reinvigoration aided the promulgation of Euro-American values and thus contributed to a certain increase in the homogeneity of values and concepts, just as gradually national times added up to form a near universal and uniform grid of time zones.

The nationalization and regionalization of time was also the result of a certain type of nineteenth- and early twentieth-century knowledge production and dissemination. Much of the knowledge and information that changed hands at international conferences and congresses in one way or another concerned the state or at least could be interpreted and made to serve the state. Nineteenth-century states discovered populations and territories as quantifiable and measurable assets and liabilities. The generation of numerical knowledge for the valuation and control of populations therefore became a prime effort.[10] Many academic disciplines and other communities of experts that participated in the international exchange of knowledge through journals, academies, and conferences traded precisely in the kind of information governments yearned for. Uniform time, although endowed with more multilayered qualities than the exegeses of railway operators and administrators could suggest, was no different in this regard. The circulation of knowledge and practices was deployed seamlessly to invigorate the process of state- and nation-building.

Viewed from the late nineteenth and early twentieth century, then, the contention that globalization and interconnectedness spell the "end of the nation-state era" is misguided. Historically, European nation-states appear to have perused the centrifugal pushes and pulls of border-crossing flows and connections quite successfully.[11] It was empires—or rather (at least initially), the old multiethnic landed empires on the Eurasian continent—that did not survive and collapsed in one form or another. Empires could not muster the resources that nation-states mobilized to harness the forces of globalization. And the tendency for knowledge and ideas to be appropriated on national terms and endowed with national functions did not work well for empires. Arab intellectuals who, around 1900, advocated time management as a tool for "national" self-improvement still did so within a framework of Ottoman rule over the Arab provinces of the empire. But the belief in an "Arab nation" and an "Islamic civilization" that stood to be strengthened in this way was a harbinger of a Syrian and Muslim identity that would very soon clash with the Ottoman (and thus, non-Arab) hold on the region in different ways and eventually contribute to the empire's disintegration, even though the Ottoman sultan sought for a while to harness the forces of Pan-Islamism to prop up the empire.[12] While empires slowly disintegrated, many cities on the other hand appeared invigorated intellectually and economically by their status as entry points and hubs for a wider region. At the same time, integration into global networks of thought and economic exchange exacerbated the divisions between these cities and their hinterlands as new trades crowded out old ones and as some of the experiments with self-strengthening and Westernization appeared too daring to more conservative minds. Yet another argument for attention to scale and differentiation between political entities, then, is that being exposed to networks of flows and connections carried strikingly different consequences for nation-states, colonies, empires, cities, and international organizations. Time reform certainly tasted differently when contemplated from the point of view of different geopolitical contexts due to the diverging interpretations and implementations of reformist thought and practices such viewpoints invited. Yet more so, such varying interpretations and applications in turn created conditions on the ground that often augmented the differences between these political entities even further.

• • • •

TIME REFORM reached its zenith roughly between 1908 and 1913. Chambers of commerce drafted resolutions on calendar reform and labored to get governments on board. An Ottoman commission discussed replacing "Arabic" or "Turkish" time officially with European time. Jubran Massuh, the journalist for *Lisan al-Hal,* wrote about his experience with "pretending to be European" and the subsequent spectacle of time-wasting he observed among the well-heeled and well-connected echelons of Levantine urban society. France adopted Greenwich time at home and throughout its empire while hosting two international conferences on telegraphic time signal distribution. An international bureau for timekeeping was set up in Paris. A sheikh in Mecca lectured on the meaning of time as fate. Maʿruf al-Rusafi, the renowned Iraqi poet, eternalized the Arab Islamic fascination with the passing of time in a poem about a clock, and about so much more.

Robert Pearce, the British MP, brought bills on calendar reform and daylight saving before the parliament. The British government appointed a commission to study summer time. The Swiss parliament and various local German bodies considered the stabilization of Easter and, together with Belgian diplomats, sought to get the Vatican on board with calendar reform. Muslim legal scholars grew worried about the diverging practices in determining the start of the Islamic lunar months; a controversy about using the telegraph in reporting the sighting of the new moon spurred a whole genre of legal thought and writing about the use of technology in the timing of religious rituals. Daylight saving associations sprang up in the Anglo-American world. The Scientific Congress of the Pan-American Union encouraged the adoption of time zones in the southwestern hemisphere. German colonial officials began to care about the application of mean times in dependent territories. Hindus and Parsis pondered the reform of their calendar. The list could go on.

Global and international histories offer the advantage of placing seemingly unique Euro-American (or for that matter, South Asian or Arab-Islamic) stories next to similar developments in other parts of the world. Both sides come into sharper relief in such a perspective. While Robert Pearce and his British contemporaries were fretting over the "waste" of daylight and time for useful activities, Arab reformers in the Levant and Egypt reprimanded fellow Easterners to spend time wisely and to manage time for acquiring useful knowledge and skills more effectively. And Ahmad Shakir's uniform Islamic calendar has to be put in conversation with proposals for a uniform world calendar as discussed by chambers of commerce, the League of Nations, and the many individual reform activists involved with the movement. Mediators and communicators of ideas were often bound together by print and communication into intellectual clusters that defied political units such as nations and regions as well as international, regional, national, and local scales altogether.

It is difficult to disentangle the mind-boggling simultaneity of these activities and to clearly identify what sustained it. At times it is possible to discern direct trajectories of transmission between single reform efforts—for example, when archival correspondence reveals exchanges between Swiss, Belgian, and German calendar reformers around 1910. In other instances, contemporaries may have read or heard of reform in one place and concluded it was time to take similar steps, albeit without leaving traces of such trajectories. In yet again further cases, there may have been no connection at all. Nineteenth-century globalization inspired people to think and talk about similar time-related questions even in the absence of direct prompts and exchanges.

Railways and telegraphs, steam and electricity certainly underpinned globalization. But for interconnectedness to be sustainable, it required people who thought and wrote about the world as a global space. Many of the individuals who nurtured globalization in such a way were journalists, publicists, and editors, public intellectuals who aptly made use of various forms of print to spread their ideas. This was true for Europe and North America as well as for the non-Western world. By the late nineteenth century, societies in the Levant, British India, and East Asia maintained a rich output of print products. As is well known, newspapers, journals, and the mass or "penny" press were important transmission belts for news and ideas in Europe in these years.[13] What is less often acknowledged is the centrality of pamphlets as a form of print culture. The format of the pamphlet is closely associated with the early modern period and the history of Enlightenment and the French and American Revolutions in particular; hence it is not commonly viewed as an important feature in the dissemination of ideas in the late nineteenth and early twentieth century, but it should be. Innovations in paper production—above all, the move away from deriving fibers from pulped rags to pulped wood instead—dramatically lowered the price of paper from the mid-nineteenth century. Even though ideas about intellectual property were beginning to form, copyright remained virtually unenforceable.[14]

The absence of serious legal hurdles, together with declining paper prices, enabled ordinary middle-class men (and fewer women) to put their thoughts into print, to advertise their products, to reprint articles published in the press or in journals, and to issue and disperse a fatwa. Government archives are littered with these little brochures, often self-published, as are libraries with holdings from the period. Their authors distributed them wherever they could, mostly unsolicited. Pamphlets and weak copyright enforcement also allowed authors to compile "best of" collages of famous literary and other works. John Murray, the Scottish publisher of many best sellers, had a story to tell about the many unlicensed and excerpted versions of his books that mushroomed in pamphlet and booklet form all over the world. Pamphlets enabled the Chinese calendar reformer to mail a hundred exemplars of his reform scheme to the League of Nations and the scholar of Islamic law to get the word out about his take on telegraphy and the lunar month. The pamphlet as a historical format was well and alive in nineteenth-century globalization. On the global level, Benedict Anderson's notion of "print-capitalism" rested on pamphlets and informal offprints just as much as on books and newspapers. Pamphlets and informal offprints or unlicensed translations and reprints did not have the same effect that Anderson postulated for newspapers, the creation of a community united in simultaneity through the act of opening the daily newspaper of the same date. But these print products nevertheless helped forge communities of thought and clusters of intellectual production beyond the nation-state, while at the same time communicating and transmitting views and opinions on time to other parts of the world.

Yet the most important factor in creating a world simultaneously preoccupied with time was less organizational and material and, rather, a matter of time's foundational quality for globalization around 1900. Time manifested itself in such variegated forms around the turn of the twentieth century because time was what helped constitute the global. Not by coincidence, evolutionary lineages, technological grids, and "modern" and "archaic" times were strongly indicative and constitutive of racial, political, social, and economic hierarchies and power relations. Time in its different guises was an underlying discursive, institutional, and technological metric that produced and permitted the emergence of a global consciousness.

Time established what appeared to be relations of commensurability and comparability by situating oneself and others in time. Nationwide mean times facilitated the calculation of time differences within regions and between neighboring countries, thus demarcating geopolitical space and, occasionally, spheres of influence. Social time created divisions of labor among different socioeconomic groups within a country by differentiating ever more specifically between occupational groups and their "times": workers, office clerks, middle-class professionals, and the "leisure" classes. Such differentiations were easily extended to the global level where they classified peoples and societies in a global division of labor. It is very telling in this regard that E. P. Thompson himself, in his famous 1967 article on time discipline, still uses select examples drawn from anthropological and ethnographic observations, such as Evans-Pritchard's writings on the Nuer, Bourdieu on the Algerian Kabyle people, and an account from the Aran Islands off the west coast of Ireland, to explain what he means by task orientation—the time of "primitive peoples," as he put it.[15]

Calendar reform and a heightened interest in calendar time in a broad range of societies were sparked by a similar comparative mind seeking to facilitate and simplify the calculation, conversion, and comparison of calendars and eras, and hence to position oneself in time and space. Among Muslim scholars, it was not so much competitive or condescending comparison that drove calendar reform but the question of how universal religious time could be in a world that was shrinking, and thus whether in a world of nations, regions, and their varying latitudes and longitudes, it was possible to imagine the Muslim community, the 'umma, as global. In his still seminal work on nationalism, Benedict Anderson argued that the simultaneity of reading a newspaper of the same date in a multitude of unrelated and distant locations "made it possible for rapidly growing numbers of people to think about themselves, and to relate themselves to others." In his famous argument, this is what contributed to the creation of "imagined communities" of nationality.[16] Anderson missed the twin nature of nationalism and nineteenth-century globalization. Simultaneity, invoked by new means of transportation and communication as much as by widely circulating pamphlets and journals within countries, inspired societies of the nineteenth century to imagine the global as much as the national. In fact, a central part of imagining the nation was to envision the nation in the world. Globalization and nationalism were intertwined in time. It is this fundamental, underlying quality of transformed time as a matrix within which to conceive of the global that makes time so central for understanding the nineteenth- and early twentieth-century world. Time talk emerged and peaked while the globe was falling under the sway of Western imperialism. Time was never neutral; there was never but one objective time that gradually came to be applied throughout the world; time reform was never merely about setting an often-dizzying array of clocks to simultaneously strike the same hour. Time and the temporalization of difference helped Europeans and, in response, non-Westerners make sense of the interconnected and heterogeneous world of the age of empire.

# Archives and Repositories

Académie Libanaise des Beaux-Arts ALBA, collection of letters and documents pertaining to Yussuf Aftimus, Beirut

American University of Beirut Jafet Library Archives and Special Collections, Beirut

Archives de la Compagnie de Jésus à Beyrouth, Beirut

Archives de la Compagnie du Port, des Quais et Entrepôts de Beyrouth, Beirut

Archives de Paris, Paris

Archives du Ministère des Affaires Etrangères, Paris–La Courneuve

Archives Nationales Françaises, Paris

British Library Asia, Pacific, and Africa Collections, India Office Records, London

British National Archives, Kew, Richmond

British Telecom Archives, London

Cambridge University Library, Royal Greenwich Observatory Archives, Cambridge

German Federal Archives, Berlin

League of Nations Archives, Geneva

Library of Congress Manuscript Division, Records of the World Calendar Association, Washington, DC

National Archives of Namibia, Windhoek

National Library of Scotland, Charles Murray Archives, Edinburgh

Royal Mail Archives, London

Smithsonian Institution National Museum of American History Archives, Washington, DC

Tanzania National Archives, Dar es Salaam

# Notes

## Abbreviations

| | |
|---|---|
| *AHR* | *American Historical Review* |
| AN | Archives Nationales Françaises |
| AP | Archives de Paris |
| AUB | American University of Beirut Archives and Special Collections, Jafet Library |
| BArch | German Federal Archives |
| BL APAC IOR | British Library Asia, Pacific, and Africa Collections, India Office Records |
| BNA | British National Archives |
| BTA | British Telecom Archives |
| HC Deb | House of Commons Debates |
| HL Deb | House of Lords Debates |
| LNA CTS | League of Nations Archives, Communications and Transit Section |
| LOC WCA | Library of Congress Manuscript Division, World Calendar Association Records |
| NAN | National Archives of Namibia |
| NLS | National Library of Scotland, John Murray Archives |
| RGO | Cambridge University Library, Royal Greenwich Observatory Archives |
| RMA | Royal Mail Archives |
| TNA | Tanzania National Archives |

## Introduction

1. German Federal Archives, Berlin Branch (hereafter BArch) R3001/7863, Reichstag, 90th session, March 16, 1891, 2092; Ian R. Bartky, *One Time Fits All: The Quest for Global Uniformity* (Stanford, CA: Stanford University Press, 2007), 126–127.
2. Arden Bucholz, *Moltke and the German Wars, 1864–1871* (New York: Palgrave, 2001); Allan Mitchell, *The Great Train Race: Railways and the Franco-German Rivalry, 1815–1914* (New York: Berghahn Books, 2000).
3. BArch R3001/7863, Reichstag, 90th session, March 16, 1891, 2092.
4. BArch R901/37725, Draft of a law regarding the introduction of a uniform definition of time, *Justification*, 4.
5. *International Conference Held at Washington for the Purpose of Fixing a Prime Meridian and a Universal Day, October 1884, Protocols of the Proceedings* (Washington, DC: Gibson Bros., 1884).
6. Karl Marx and Friedrich Engels, *Manifesto of the Communist Party, Authorized English Translation Edited and Annotated by Frederick Engels* (New York: New York Labor News Co., 1908), 12; Charles Bright and Michael Geyer, "Regimes of World Order: Global Integration and the Production of Difference in Twentieth-Century World History," in *Interactions: Transregional Perspectives on World History,* ed. Jerry Bentley, Renate Bridenthal, and Anand Yang (Honolulu: University of Hawai'i Press, 2005), 202–238, here 219; Patrick O'Brien, ed., *Railways and the Economic Development in Western Europe, 1830–1914* (New York: St. Martin's Press, 1983); Roland Wenzlhuemer, *Connecting the Nineteenth-Century World: The Telegraph and Globalization* (Cambridge: Cambridge University Press, 2013); David Hochfelder, *The Telegraph in America, 1832–1920* (Baltimore: Johns Hopkins University Press, 2012); Daniel Headrick, *Invisible Weapon: Telecommunications and International Politics, 1851–1945* (New York: Oxford University Press, 1991), and for an imperial context, Daniel Headrick, *Tentacles of Progress: Technology Transfer in the Age of Imperialism* (New York: Oxford University Press, 1988).
7. For a critical view on the usefulness of "globalization" for historians, see Frederick Cooper, "Globalization," in Cooper, *Colonialism in Question* (Berkeley: University of California Press, 2005), 91–112. On definitions and periodization of globalization, see Lynn Hunt, *Writing History in the Global Era* (New York: W. W. Norton, 2014), 52; Anthony Hopkins, "Globalization—An Agenda for Historians," in *Globalization in World History,* ed. Hopkins (New York: Norton, 2002), 1–11; Adam McKeown, "Periodizing Globalization," *History Workshop Journal* 63, no. 1 (2007): 218–230; Jürgen Osterhammel, *Globalization: A Short History* (Princeton, NJ: Princeton University Press, 2005). For an economic perspective, see Kevin O'Rourke and Jeffrey Williamson, *Globalization and History: The Evolution of a Nineteenth-Century Atlantic Economy* (Cambridge, MA: Harvard University Press, 1999); Michael Bordo, Alan Taylor, and Jeffrey Williamson, eds., *Globalization in Historical Perspective* (Chicago: University of Chicago Press, 2003); see also Charles Bright and Michael

Geyer, "World History in a Global Age," *American Historical Review* 100, no. 4 (1995): 1034–1060; Charles S. Maier, "Consigning the Twentieth Century to History: Alternative Narratives for the Modern Era," *American Historical Review* 105, no. 3 (2000): 807–831. For a Middle Eastern perspective, see James Gelvin and Nile Green, eds., *Global Muslims in the Age of Steam and Print* (Berkeley: University of California Press, 2014).

8. Robert Schram, *Adria-Zeit, Separatabdruck aus der "Neuen Freien Presse"* (Vienna: Self-published, 1889), 3.
9. On "mentalities," see André Burguière, *The Annales School: An Intellectual History,* trans. Jane Marie Todd (Ithaca, NY: Cornell University Press, 2009), ch. 3; David Harvey, *The Condition of Postmodernity: An Enquiry into the Origins of Social Change* (Oxford: Blackwell, 1989), 240; Adam McKeown, *Melancholy Order: Asian Migration and the Globalization of Borders* (New York: Columbia University Press, 2008); Sebastian Conrad, *Globalization and the Nation in Imperial Germany* (Cambridge: Cambridge University Press, 2010), 50.
10. On tensions between nationalism and internationalism, see Emily Rosenberg, "Transnational Currents in a Shrinking World," in *A World Connecting, 1870–1945,* ed. Rosenberg (Cambridge, MA: Belknap Press of Harvard University Press, 2012), 815–995, here 821; Glenda Sluga, *Internationalism in the Age of Nationalism* (Philadelphia: University of Pennsylvania Press, 2013).
11. Elisabeth Crawford, "The Universe of International Science, 1880–1939," in *Solomon's House Revisited: The Organization and Institutionalization of Science,* ed. Tore Frängsmyr (Canton, MA: Science History Publications, 1990), 251–269, here 255; Elisabeth Crawford, Terry Shinn, and Sverker Sörlin, "Introduction," in *Denationalizing Science: The Contexts of International Scientific Practice,* ed. Crawford, Shinn, and Sörlin (Dordrecht: Kluwer, 1993), 1–42, here 14.
12. Durkheim's "Comparative Method," laid out in his 1895 work on methodology, is one example. Emile Durkheim, *The Rules of Sociological Method,* ed. Steven Lukes, trans. W. D. Halls (New York: Free Press, 1982), esp. ch. 6.
13. Friedrich Nietzsche, *Human, All Too Human, A Book for Free Spirits,* trans. R. J. Hollingdale (Cambridge: Cambridge University Press, 1986), 24.
14. Edmond Demolins, *A quoi tient la supériorité des Anglo-Saxons* (Paris: Firmin-Didot, 1897).
15. Stephen Kern, *The Culture of Time and Space* (Cambridge, MA: Harvard University Press, 1983).
16. See the contributions in Christoph Conrad and Sebastian Conrad, eds., *Die Nation Schreiben. Geschichtswissenschaft im internationalen Vergleich* (Göttingen: Vandenhoeck & Ruprecht, 2002); Ulrike Freitag, "Notions of Time in Arab-Islamic Historiography," *Storia della Storiografia* 28 (1995): 55–68, here 64.
17. Benedict Anderson, *Imagined Communities: Reflections on the Origin and Spread of Nationalism* (London: Verso, 1991), 26. On nations and nationalism as made and remade through border-crossing interactions, see McKeown, *Melancholy Order;* Manu Goswami, *Producing India: From Colonial Economy to*

*National Space* (Chicago: University of Chicago Press, 2004); Jordanna Bailkin, *The Afterlife of Empire* (Berkeley: University of California Press, 2012), charts the influence of decolonization on the quintessentially national institution of the welfare state.

18. Johannes Fabian, *Time and the Other: How Anthropology Makes Its Object* (New York: Columbia University Press, 2014 [1983]), 31. See also Dipesh Chakrabarty, *Provincializing Europe: Postcolonial Thought and Historical Difference* (Princeton, NJ: Princeton University Press, 2008), 8.
19. Michel Foucault, *Discipline and Punish: The Birth of the Prison,* trans. Alan Sheridan (New York: Vintage Books, 1979).
20. Jürgen Osterhammel, *The Transformation of the World: A Global History of the Nineteenth Century* (Princeton, NJ: Princeton University Press, 2014), 72.
21. Marx and Engels, *Manifesto of the Communist Party,* 13.
22. Moishe Postone, *Time, Labor, and Social Domination: A Reinterpretation of Marx's Critical Theory* (Cambridge: Cambridge University Press, 1993), 202, 211. See also Anthony Giddens, *The Consequences of Modernity* (Stanford, CA: Stanford University Press, 1990), 17.
23. E. P. Thompson, "Time, Work Discipline, and Industrial Capitalism," *Past and Present* 38 (December 1967): 56–97.
24. David Landes, *Revolution in Time: Clocks and the Making of the Modern World* (Cambridge, MA: Belknap Press of Harvard University Press, 1983); Carlo Cipolla, *Clocks and Culture, 1300–1700* (London: Collins, 1967); Jacques Le Goff, "Merchant's Time and Church's Time in the Middle Ages," in *Time, Work, and Culture in the Middle Ages,* ed. Jacques Le Goff, trans. Arthur Goldhammer (Chicago: University of Chicago Press, 1980), 29–42.
25. Derek Howse, *Greenwich Time and the Discovery of the Longitude* (Oxford: Oxford University Press, 1980), 50–53.
26. Ibid., 67–72, 198.
27. See ibid., 134, for a list of meridians in use.
28. For histories of science and technology on time reform, see Ian R. Bartky, *Selling the True Time: Nineteenth-Century Time-Keeping in America* (Stanford, CA: Stanford University Press, 2000), and Bartky, *One Time Fits All;* Peter Galison, *Einstein's Clocks, Poincarés Maps: Empires of Time* (New York: W. W. Norton, 2003); Howse, *Greenwich Time.* On time in a national perspective, see Galison on France; Jakob Messerli, *Gleimässig, pünktlich, schnell: Zeiteinteilung und Zeitgebrauch in der Schweiz im 19. Jahrhundert* (Zurich: Chronos, 1995); on the United States, see Michael O'Malley, *Keeping Watch: A History of American Time* (New York: Viking, 1990); Carlene Stephens, *On Time: How America Has Learned to Live by the Clock* (Boston: Bulfinch Press, 2002). A notable exception among historians of science and technology is Ken Alder, *The Measure of All Things: The Seven-Year Odyssey and Hidden Error that Transformed the World* (New York: Free Press, 2002), on the establishment of the meter in revolutionary France, a book based on a more mixed archival basis.
29. The exception being Japan, which had managed to claim a seat at the table of the "civilized" countries by the late nineteenth century.

30. On daylight saving, see Bartky, *One Time Fits All,* 161–183, and two books written for a broader audience: Michael Downing, *Spring Forward: The Annual Madness of Daylight Saving* (Washington, DC: Shoemaker & Hoard, 2005); David Prerau, *Seize the Daylight: The Curious and Contentious Story of Daylight Saving Time* (New York: Thunder's Mouth Press, 2005).
31. Maʿruf al-Rusafi, "Al-Saʿa," *al-Muqtabas* 3, no. 28 (1908): 279; Jamal al-Din al-Qasimi, *Irshad al-Khalq ila al-ʿAmal bi-Khabar al-Barq* (Damascus: Maṭbaʿa al-Muqtabas, 1911); Muhammad Bakhit al-Mutiʿi, *Irshad 'Ahl al-Milla ila Ithbat al-Ahilla,* ed. Hasan Ahmad Isbir (Beirut: Dar Ibn Hazm, 2000/1911). On self-strengthening and reform movements arguing for the strategic adoption of Western knowledge, see Osterhammel, *Transformation,* ch. XI.
32. Luca Carboni, "Cesare Tondini, gli anni della giovinezza: 1839–1871 (formazione, missione e primi scritti)," *Studi Barnabiti* 22 (2005): 95–195; Ebrahim Moosa, "Shaykh Aḥmad Shākir and the Adoption of a Scientifically-Based Lunar Calendar," *Islamic Law and Society* 5, no. 1 (1998): 57–89.
33. Elisabeth Achelis, *The World Calendar: Addresses and Occasional Papers Chronologically Arranged on the Progress of Calendar Reform since 1930* (New York: Putnam, 1937).
34. On the "Equation of Time" by which these variations are homogenized, see Howse, *Greenwich Time,* 38.

## 1. National Times in a Globalizing World

1. On state formation and national integration, see Geoff Eley, "State Formation, Nationalism, and Political Culture: Some Thoughts on the German Case," in *Culture, Ideology, and Politics (Festschrift for Eric Hobsbawm),* ed. Raphael Samuel and Gareth Stedman Jones (London: Routledge and Kegan Paul, 1983), 277–301.
2. See Dirk Hoerder, *Cultures in Contact: World Migrations in the Second Millennium* (Durham, NC: Duke University Press, 2002), esp. 331–404; McKeown, "Periodizing Globalization," esp. 31; McKeown, "Global Migration 1846–1940," *Journal of World History* 15, no. 2 (2004): 155–189.
3. Cornelius Torp, *Die Herausforderung der Globalisierung: Wirtschaft und Politik in Deutschland 1860–1914* (Göttingen: Vandenhoeck & Ruprecht, 2005), 27–42.
4. John MacKenzie, *Propaganda and Empire: The Manipulation of British Public Opinion, 1880–1960* (Manchester: Manchester University Press, 1984); David Ciarlo, *Advertising Empire: Race and Visual Culture in Imperial Germany* (Cambridge, MA: Harvard University Press, 2011).
5. Jane Burbank and Frederick Cooper, *Empires in World History: Power and the Politics of Difference* (Princeton, NJ: Princeton University Press, 2010), 288; Eric Hobsbawm, *The Age of Empire* 1875–1914 (London: Vintage Books, 1989), 59.
6. Mary Nolan, *The Transatlantic Century: Europe and America, 1890–2010* (Cambridge: Cambridge University Press, 2012), 11, 17; D. K. Fieldhouse, *Economics and Empire, 1830–1914* (London: Weidenfeld and Nicolson, 1973).

7. Akira Iriye, *Global Community: The Role of International Organizations in the Making of the Contemporary World* (Berkeley: University of California Press, 2002); Daniel Gorman, *The Emergence of International Society in the 1920s* (Cambridge: Cambridge University Press, 2012); Rosenberg, "Transnational Currents in a Shrinking World."
8. H. La Fontaine and P. Otlet, "La vie internationale et l'effort pour son organization," *La Vie Internationale* 1 (1912): 9–34, here 12.
9. Archives Nationales Françaises (hereafter AN) F/17/3714, Adaption d'une heure unique, Note de Sandford Fleming, "Memorandum on the Movement for Reckoning Time on a Scientific Basis, by which the greatest possible degree of Simplicity, Accuracy, and Uniformity will be obtained in all Countries throughout the World," Ottawa, November 20, 1889, 1.
10. Crawford, "Universe of International Science," 255.
11. BArch R901/63559, pamphlet by the Freie deutsche Hochstift für Wissenschaften, Künste und allgemeine Bildung in Goethes Vaterhause, Frankfurt/Main, January 1, 1864.
12. Alexis McCrossen, *Marking Modern Times: A History of Clocks, Watches, and Other Timekeepers in American Life* (Chicago: University of Chicago Press, 2013), 10.
13. G. Hammer, *Nullmeridian und Weltzeit* (Hamburg: Verlagsanstalt und Druckerei A. G., 1888), 44.
14. Bartky, *One Time Fits All,* 74.
15. BArch R 901/63559, F. Romanet du Caillaud, De l'Adoption du Méridien de Bethléem comme premier Méridien universel, Lettre à Monsieur le Président de la société de géographie de Paris, Limoges, September 6, 1884.
16. AN F/17/3714, Adaption d'une heure unique, Italie; Resolutions by the International Geodesic Association Concerning the Unification of Longitudes and Time, Bureau central de l'Association Géodésique Internationale, n.d.
17. Bartky, *One Time Fits All,* 72.
18. AN F/17/3714, Conférence internationale de Washington pour l'établissement d'un méridien initial, letter, Ministry of Foreign Affairs to Ministry of Public Instruction, January 26, 1884; Circular, Department of State, Washington, DC, December 1, 1883; BArch R901/63559, Legation of the United States to German Foreign Ministry, Berlin, November 9, 1882; Bartky, *One Time Fits All,* 72.
19. Theodore M. Porter, *Trust in Numbers: The Pursuit of Objectivity in Science and Public Life* (Princeton, NJ: Princeton University Press, 1995); Ian Hacking, *The Taming of Chance* (New York: Cambridge University Press, 1990); and the contributions in M. Norton Wise, ed., *The Values of Precision* (Princeton, NJ: Princeton University Press, 1995), esp. Simon Schaffer, "Accurate Measurement Is an English Science," 135–172, and M. Norton Wise, "Precision: Agent of Unity and Product of Agreement, Part II—The Age of Steam and Telegraphy," 222–236; and Simon Schaffer, "Late Victorian Metrology and Its Instrumentation: A Manufactory of Ohms," in *Invisible Connections: Instruments, Institutions, and Science,* ed. Robert Bud and Susan Cozzens (Bellingham, WA: SPIE, 1992), 23–56, here 25.

20. Schram, *Adria-Zeit,* 4, 12; W. d Nordling, "Les derniers progrès de l'unification de l'heure," *Revue scientifique* 30, no. 1 (1893): 774–777, here 775; Cambridge University Library, Royal Greenwich Observatory Archives (hereafter RGO) 7/146, clipping "La Belgique et l'heure de Greenwich," *Le mouvement géographique,* May 1, 1892, 1.
21. Friedrich Naumann, *Mitteleuropa* (Berlin: Georg Reimer, 1915), 3.
22. Jürgen Elvert, *Mitteleuropa! Deutsche Pläne zur europäischen Neuordnung (1918–1945)* (Stuttgart: Franz Steiner Verlag, 1999), 38.
23. Schmitt was justifying the Nazi annexations in Central Europe during the second half of the 1930s. See Carl Schmitt, *Völkerrechtliche Grossraumordnung mit Interventionsverbot für raumfremde Mächte. Ein Beitrag zum Reichsbegriff im Völkerrecht* (Berlin: Duncker & Humblot, 1991); Schmitt mentions Naumann only in passing (p. 12) but counts him among the precursors for the kind of concept he is proposing. See also Felix Blindow, *Carl Schmitts Reichsordnung. Strategien für einen europäischen Grossraum* (Berlin: Akademie Verlag, 1999).
24. Naumann, *Mitteleuropa,* 166.
25. Max Reithoffer, "Ein elektrisches Zentraluhrensystem fur Wien, Vortrag, gehalten den 22. Februar 1911," *Schriften des Vereines zur Verbreitung Naturwissenschaftlicher Kenntnisse in Wien* 51 (1910/1911): 439–456, here 440.
26. See Archives de Paris (hereafter AP), VONC 20, draft letter, Direction of Public Works, May 1880; letter, Mayoral Office of the 17th District, Paris, June 2, 1883; Galison, *Einstein's Clocks,* 93–97.
27. AN F/17/3714, Adoption of a single time, Denmark; Decree about the renewed determination of time, January 5, 1894; Adoption of a single time, Norway; letter, Ministry of Foreign Affairs, Ministry of Public Instruction, January 9, 1895; Adoption of a single time, Spain, letter, Ministry of Foreign Affairs to Ministry of Public Instruction, August 16, 1900; BArch R901/37725, Swiss Legation in Berlin, received June 5, 1894; BArch R3001/7863, Berlin, September 13, 1893.
28. Oliver Zimmer, *Remaking the Rhythms of Life: German Communities in the Age of the Nation-State* (Oxford: Oxford University Press, 2013), 5; Siegfried Weichlein, *Nation und Region: Integrationsprozesse im Bismarckreich* (Düsseldorf: Droste Verlag, 2004); see also Abigail Green, *Fatherlands: State-Building and Nationhood in Nineteenth-Century Germany* (Oxford: Oxford University Press, 2001). On interactions between municipalities and the nation-state in different case studies, see Zimmer, *Remaking*; Laurent Brassart, Jean-Pierre Jessenne, and Nadine Vivier, eds., *Clochemerle our république villageoise? La conduit municipal des affaires villageoises en Europe du XVIIIè au XXè siècle* (Villeneuve d'Ascq: Presses Universitaires du Septentrion, 2012).
29. Schram, *Adria-Zeit,* 4.
30. BArch R901/15611, Copy, letter by Ministry of Public Works, Berlin, November 9, 1891, 1–2.
31. Weichlein, *Nation und Region,* 104.
32. BArch R901/15609, memorandum, Berlin, December 1889, 9.

33. BArch R3001/7863, Opinion by the Vice President of the Ministry of State, von Boetticher, regarding the uniform measure of time, Berlin, February 15, 1892; see also letter, Secretary of State for the Interior to Ministry of Justice, Berlin, February 15, 1892; copy of memorandum, Berlin, December 1889.
34. BArch R 901/15609, memorandum, Berlin, December 1889, 11.
35. BArch R 901/37725, Opinion, Prussian Ministry of Spiritual, Educational, and Medical Matters concerning the legal-imperial adoption of a uniform time for the entire civil life in Germany, Berlin, December 21, 1891. See also Wilhelm Förster, *Über Zeitmessung und Zeitregelung* (Leipzig: Johann Ambrosius Barth, 1909), 92.
36. BArch R 901/37725, Reichstag, 28th session, Monday, January 23, 1893, 634.
37. Quoted after Weichlein, *Nation und Region,* 39.
38. Ibid., 101–102.
39. Nordling, "Les derniers progrès," 775.
40. Bartky, *One Time Fits All,* 126.
41. BArch R3001/7863, Imperial Office of Railways to railway administrations, March 19, 1893.
42. BArch R901/63559, letter, Ministry of Spiritual, Educational, and Medical Affairs, Berlin, April 2, 1883.
43. BArch R901/63559, Instruction to Supervisor of Constructions Hinckeldeyn regarding the consultations of the international conference to be opened in Washington for the introduction of a uniform Meridian and a uniform time, n.d. (Karl Hinckeldeyn was an architect who, in 1884, served as technical attaché to the German embassy in Washington, DC.) See also ibid., Note Regarding the Meridian Conference in Washington, Berlin, October 14, 1884.
44. BArch R901/37725, Reichstag, 28th session, Monday, January 23, 1893, 634. See also Order issued by the Ministry of Church and Schooling Affairs in Wurtemberg: Verfügung des Ministeriums des Kirchen- und Schulwesens, betreffend die Ordnung der Unterrichtszeit in den Volksschulen und den kleineren Gelehrten- und Realschulen infolge der Einführung der sogenannten mitteleuropäischen Einheitszeit, Stuttgart, February 23, 1892.
45. RGO 7/146, newspaper offprint, Robert Schram, "Die Zeitreform in Belgien," *Deutsche Zeitung,* May 31, 1892.
46. BArch R3001/7863, letter, German Embassy in The Hague to German Ministry of Foreign Affairs, June 4, 1896.
47. BArch R3001/7863, letter, German Embassy in The Hague to Ministry of Foreign Affairs, June 11, 1897.
48. BArch R901/37725, Reichstag, 28th session, Monday, January 23, 1893, 634.
49. BArch R901/15612, clipping "Local Time and Mean Time Once Again," *Kölnische Zeitung,* May 19, 1892.
50. On the French quest for a metric system, see Kenneth Alder, "A Revolution to Measure: The Political Economy of the Metric System in France," in Wise, *The Values of Precision,* 38–71, and Paul Smith, "La division décimale du jour: l'heure qu'il n'est pas," in *Génèse et diffusion du système métrique: actes du colloque la naissance du système métrique, URA-CNRS 1013 et 1252, Musée*

*national des techniques, CNAM, 20–21 octobre 1989,* ed. Bernard Garnier and Jean-Claude Hocquet (Caen: Editions diffusion du Lys, 1990), 123–135.

51. See Galison, *Einstein's Clocks,* on the bureau and especially its triangulation activities; Matthew Shaw, *Time and the French Revolution: The Republican Calendar, 1789-Year XIV* (Woodbridge, UK: Royal Historical Society, 2011). See also Noah Shusterman, *Religion and the Politics of Time: Holidays in France from Louis XIV through Napoleon* (Washington, DC: Catholic University of America Press, 2010).
52. AN F/17/3714, Conférence internationale de Washington pour l'établissement d'un méridien initial, Report, A. Lefaire, New York, October 28, 1884.
53. F. A. Forel, "L'heure nationale Française," *Revue scientifique* 41, no. 1 (1888): 806–809, here 807.
54. M. L. Lossier, "La fabrication des montres et l'enseignement de l'horlogerie à Besançon," *Revue Scientifique* 28, no. 1 (1891): 196–203, here 196. R. L. Reverchon, "Les observatories chronométriques," *Revue Scientifique* 33, no. 2 (1896): 655–658.
55. E. Cugnin, "L'heure et la longitude décimale et universelle," *Revue Scientifique* 40, no. 2 (1903): 193–203, here 193.
56. Reverchon, "Les observatories chronométriques," 655–658.
57. Cugnin, "L'heure et la longitude,"193.
58. AN F/17/3713, Unification de l'heure en France et en Algérie, Commission de Géodesie, Extrait du registre des Délibérations du Conseil Municipal de Langres, November 6, 1888.
59. AN F/17/3713, Unification de l'heure en France et en Algérie, Commission de Géodesie, Société de géographie commerciale de Bordeaux to Ministry of Public Instruction, January 29, 1890. See Robert Fox, "The Savant Confronts His Peers: Scientific Societies in France, 1815–1914," in *The Organization of Science and Technology in France, 1804–1914,* ed. Robert Fox and Donald Weisz (Cambridge: Cambridge University Press, 1980), 241–282.
60. AN F/17/3713, Unification de l'heure en France et en Algérie, Commission de Géodesie, Conseil Général, Côte d'Or, First Ordinary Session of 1890, session of April 18, 1890.
61. AN F/17/3713, Unification de l'heure en France et en Algérie, Commission de Géodesie, Ville de Lyon, Extract from the Registry of Deliberations of the Municipal Council, December 17, 1889; Bureau des Longitudes Ministry of Public Instruction, January 6, 1890; Guillaume Bigourdan, *Le jour et ses divisions. Les fuseaux horaires et les conferences internationals de l'heure en 1912 et 1913* (Paris: Gauthier-Villars, 1914), B. 43.
62. AN F/17/3713, Bureau des Longitudes, No. 2072, Chambre des Députés, Proposition de loi, Session of October 27, 1896; see also Charles Lallemand, *L'unification internationale des heures et le système des fuseaux horaires* (Paris: Bureau de la Revue Scientifique, 1897), 9; Bigourdan, *Le jour et ses divisions,* 59.
63. "La suppression du méridien de Paris," *Le Petit Parisien,* September 28, 1898.
64. AN F/17/2921, Commission de décimalisation, Ministère de l'instruction publique, des beaux-arts et des cultes, Note pour Monsieur le Ministre, Paris,

November 29, 1905. On debates among mathematicians and other scientists on decimal time, see Galison, *Einstein's Clocks,* 162–174.

65. AN F/17/3713, Bureau des Longitudes, No. 2326, Chambre des Députés, Proposition de loi, Session of March 8, 1897; Bigourdan, *Le jour et ses divisions,* 61.
66. AN F/17/3713, Bureau des Longitudes, Travaux Géodétiques, No. 3039, Chambre des Députés, Proposition de loi, Session of February 16, 1898.
67. Bigourdan, *Le jour et ses divisions,* 67; "A propos de la nouvelle heure légale," *Le Petit Journal,* March 10, 1911, 1; "L'unification de l'heure," *Le Petit Journal,* February 19, 1911, 1.
68. David Cahan, *An Institute for an Empire: The Physikalisch-Technische Reichsanstalt, 1871–1918* (Cambridge: Cambridge University Press, 1989).
69. On science, the state, and colonial expansion, see Lewis Pyenson, *Civilizing Mission: Exact Sciences and French Overseas Expansion, 1830–1940* (Baltimore: Johns Hopkins University Press, 1993), 15. On the British context, see D. Graham Burnett, *Masters of All They Surveyed: Exploration, Geography, and a British El Dorado* (Chicago: University of Chicago Press, 2000).
70. On nineteenth-century observatories, see David Aubin, Charlotte Bigg, and H. Otto Sibum, eds., *Heavens on Earth: Observatories and Astronomy in Nineteenth-Century Science and Culture* (Durham, NC: Duke University Press, 2010).
71. AN F/17/13577, Conférence internationale de l'heure, Guillaume Bigourdan, head of Bureau of Longitudes, Rapport sur les travaux de la conférence internationale pour l'unification de l'heure par radiotélégraphie, November 1912; Bigourdan, letter to Ministry of Public Instruction, February 1, 1912; Paul Jegou, "Determination de l'heure et mesure des différences de longitude au moyen des signaux horaires et pendulaires Hertziens," *Revue scientifique* 49, no. 2 (1911): 37–43, esp. 37. See also G. Ferrié, "La télégraphie sans fil et le problème de l'heure," *Revue scientifique* 51, no. 2 (1913): 70–75, here 71.
72. Ibid. See also AN F/17/13577, Projet de conférence rélative au service international d'indications de l'heure, Note, Mai 6, 1912; Note, Paris, May 6, 1912. For a comparison of French and German facilities and plans for an international time service, see also ibid., Avant-Projet d'organisation d'un service international de l'heure.
73. AN F/17/13577, Paris, Conference destinée à la signature de la convention internationale (lundi 20 octobre 1913), Secretary of the Academy of the Sciences to Ministry of Public Instruction, November 17, 1913.
74. BArch R1001/6190 Ministry of Foreign Affairs to French Ambassador, Berlin, July 31, 1913. See also ibid., meeting of ministerial representatives on the question of the time convention, protocol, October 7, 1913; AN F/17/13577, Conférence destinée à la signature de la convention internationale (lundi 20 octobre 1913), Procès verbaux de la conférence internationale de l'heure (1913), Séance du 20 Octobre 1913, 4.

75. BArch R1001/6190, October 7, 1913.
76. AN F/17/13577, Conférence destinée à la signature de la convention internationale (lundi 20 octobre 1913), Procès verbaux de la conférence internationale de l'heure (1913), Séance du 20 Octobre 1913, 5.
77. AN F/17/13577, Conférence destinée à la signature de la convention internationale (lundi 20 octobre 1913), Note, Secretary of State, Washington, DC, July 23, 1913.
78. AN F/17/13577, Conférence internationale de l'heure, letter, Bureau des Longitudes to Ministry of Public Instruction, November 13, 1912. AN F/17/13577, Conférence destinée à la signature de la convention internationale (lundi 20 octobre 1913), Convention pour la création d'une association internationale de l'heure, October 25, 1913, and Statutes de l'association internationale de l'heure; letter, Bureau des Longitudes to Ministry of Public Instruction, November 13, 1912; Bigourdan, *Le jour et ses divisions,* 9.
79. BArch R4701/207, letter, Minister of Public Works to Foreign Office, Berlin, October 28, 1919.
80. "Städtische Nachrichten: Westeuropäische Zeit für den Eisenbahnverkehr," *Kölnische Zeitung,* October 3, 1920, 1–2.
81. BArch R4701/207, letter, Foreign Office to Ministry of Transportation, Berlin, September 1, 1920.

## 2. Saving Social Time

1. Postone, *Time, Labor, and Social Domination,* sees an interest in improved technology and a demand for clocks as propelled by the social conditions of early capitalism.
2. Thompson, "Time, Work Discipline, and Industrial Capitalism," 60; Eviatar Zerubavel, "The Standardization of Time: A Sociohistorical Perspective," *American Journal of Sociology* 88, no. 1 (1982): 1–23, here 19; Postone, *Time, Labor, and Social Domination*, 211. Marx himself discussed time in a number of contexts (e.g., surplus value, working day) and, most pertinent here, in his characterization of the commodity. A commodity only obtains value because abstract human labor went into producing it. "How, then, is the magnitude of this value to be measured? By means of the quantity of the 'value-forming substance,' the labor, contained in the article. This quantity is measured by its duration, and the labor-time is itself measured on the particular scale of hours, days etc." Karl Marx, *Capital: A Critique of Political Economy*, vol. 1 (London: Penguin, 1990), 129. Time is, as Postone would put it, no longer "measured by labor," but it now "measures labor." Postone, *Time, Labor, and Social Domination*, 216. On the early modern and medieval context, see Le Goff, "Merchant's Time," 29–43.
3. O'Malley, *Keeping Watch,* ix.
4. Thompson, "Time, Work Discipline, and Industrial Capitalism," 80.
5. William Willett, *The Waste of Daylight: With an Account of the Progress of the Daylight Saving Bill* (London: Self-published, 1907), 3, 4.

6. Anson Rabinbach, *The Human Motor: Energy, Fatigue, and the Origins of Modernity* (New York: Basic Books, 1990); Patrick Joyce, "Work," in *The Cambridge Social History of Britain, 1750–1950,* vol. 2: *People and Their Environment,* ed. F. M. L. Thompson (Cambridge: Cambridge University Press, 1990), 131–194, here 132.
7. Hobsbawm, *Age of Empire,* 53.
8. On sociotemporal rhythms, see Eviatar Zerubavel, *Hidden Rhythms: Schedules and Calendars in Social Life* (Chicago: University of Chicago Press, 1981), esp. 2; Willett, *Waste of Daylight,* 4; British National Archives, Kew (hereafter BNA), HO 45/19959, Summer Time Committee, Report of the Committee Appointed by the Secretary of State for the Home Department, London, 1917, 20; BNA HO 45/10548/162178, Victoria, Abridged Report from the Select Committee upon the Saving of Daylight Together with the Minutes of Evidence, 1909, 1; "The United States. Daylight Saving in Cincinnati," *Times of London,* July 3, 1909, 5. On daylight saving in the United States, see O'Malley, *Keeping Watch,* ch. 6.
9. O'Malley, *Keeping Watch,* ch. 4.
10. BNA HO 45/10548/162178, The Daylight Saving Bill, Deputation to Reginald McKenna, Secretary of State for the Home Department, March 24, 1914, 1, 3, 20.
11. Ibid., 3. On the idea of wartime since World War II, see Mary Dudziak, *Wartime: An Idea, Its History, Its Consequences* (Oxford: Oxford University Press, 2012).
12. BNA HO 45/11626, unnamed memorandum, received at Home Office November 6, 1923; "At Last," *Times of London,* July 18, 1925, 13.
13. Royal Mail Archives (hereafter RMA), POST 31/11B, Royal Scottish Geographical Society, memorandum, Notes on Standard Time, Edinburgh, July 24, 1898; letter, General Post Office to Royal Scottish Geographical Society, London, August 6, 1898.
14. House of Commons Debates (hereafter HC Deb), May 25, 1911, vol. 26, c. 580; see also HC Deb, June 20, 1911, vol. 27, c. 220; HC Deb, July 10, 1912, vol. 12, c. 390.
15. HC Deb, March 4, 1912, vol. 11, c. 288.
16. House of Lords Debates (hereafter HL Deb), June 24, 1912, vol. 12, c. 132. See also HL Deb, July 24, 1912, vol. 12, cc. 696–698.
17. HL Deb, July 10, 1912, vol. 12, c. 387.
18. HC Deb, August 1, 1916, vol. 85, c. 73.
19. RMA POST 33/1492B, Time Act; BNA HO 45/19959, Report on the working of summer time in Ireland in 1917, n.d.; HC Deb, April 19, 1917, vol. 92, c. 1873; HO 45/10811/312364, letter, Irish Office to Home Office, May 16, 1916; HO 45/10811/312364, Time (Act) Ireland, September 12, 1916; RMA POST 33/1492B, Time Act.
20. BArch R3001/7864, Aufzeichnung über die Sommerzeit, March 20, 1922.
21. BArch R3001/7864, letter, Ministry of the Interior to Prussian Minister of Sciences, Art, Education, Berlin, March 29, 1924; Nationalversammlung, 35th Session, April 11, 1919, 975.

22. BNA HO 45/10811/312364, Rapport par M. Guilloteaux, presented to the French Senate Session of June 6, 1916, 2.
23. Ibid., 11.
24. BNA HO 45/11077/411632, Extract from the "Journal Officiel," March 15, 1922. During World War I, almost all European countries observed summer time: besides Britain, Ireland, France, and Germany, the list included Italy, the Netherlands, Norway, Austria, Switzerland, Denmark, Sweden, and Spain. See HO 45/11940, letter, Home Office to the Labor Research Department, March 14, 1924; see BNA HO 45/19959, Summer Time Committee, Report of the Committee Appointed by the Secretary of State for the Home Department, London, 1917, 19.
25. BNA HO 45/11940, Summer Time, memorandum by the Home Secretary, March 4, 1924, 4; BNA HO 45/11077/411632, notes of an Interdepartmental Conference held at the Home Office on the 27th April, 1921, on the Desirability of International Uniformity of Summer Time; BNA HO 45/11940, note of a Conference held at Home Office on November 22, 1921, 1, 3.
26. BNA HO 45/11940, letter, Advisory and Technical Committee to Home Office, London, September 12, 1923; letter to the Cabinet Office, August 3, 1922, as well as letter, Secretary General of the League of Nations to British Prime Minister, Geneva, July 7, 1922; BNA HO 45/11626, Summer Time—Notes of Deputation from the Early Closing Association, n.d., 2, 8; EU Directive 2000/84/EC, http://eur-lex.europa.eu/LexUriServ/LexUriServ.do?uri=CELEX:32000L0084:EN:NOT.
27. BNA HO 45/10548/162178, "The Waste of Daylight (to the Editor of the 'Spectator')," *The Spectator*, August 24, 1907; HC Deb, March 5, 1909, vol. 1, c. 1744, 1761.
28. BNA HO 45/10548/162178, memorandum, The Daylight Saving Bill, March 24, 1914, 8. See also Brad Beaven, *Leisure, Citizenship, and Working-Class Men in Britain, 1850–1945* (New York: Palgrave, 2005), 34–36 and 66–72; BNA HO 45/11626, Summer Time—Notes of Deputation from the Early Closing Association, n.d., 4; Gary Cross, *The Quest for Time: The Reduction of Work in Britain and France, 1840–1940* (Berkeley: University of California Press, 1989), 8.
29. Hugh Cunningham, *Leisure in the Industrial Revolution* (London: Croom Helm, 1980), 282–283; according to Hobsbawm, real wages among workers in fact declined between roughly the 1890s and 1914, but the steep decline in prices that came with the economic depression of the 1870s and 1880s nevertheless left workers with more to spend beyond subsistence than before. Hobsbawm, *Age of Empire*, 48–49.
30. Cunningham, *Leisure*, 288, 294; also Hugh Cunningham, *Time, Work and Leisure: Life Changes in England Since 1700* (Manchester: Manchester University Press, 2014); Hobsbawm, *Age of Empire*, 174.
31. Beaven, *Leisure, Citizenship*, 16.
32. BNA HO 45/11940, Parliamentary Debates, House of Commons, Standing Committee C, Summer Time Bill, Official Report, March 24, 1925, c. 21.
33. BNA HO 45/11626, Summer Time—Notes of Deputation from the Early Closing Association, n.d., 5–6.

34. Ibid., 7.
35. BNA HO 45/10548/162178, memorandum, The Daylight Saving Bill, March 24, 1914, 4; Willett, *Waste of Daylight*, 5.
36. BNA HO 45/19959, Summer Time Committee, Report of the Committee Appointed by the Secretary of State for the Home Department, London, 1917, 4, 9.
37. BNA HO 45/10548/162178, letter, James to Home Office, Cornwall, May 5, 1916.
38. BNA HO 45/19959, Summer Time Committee, Report of the Committee Appointed by the Secretary of State for the Home Department, London, 1917, 15; Report and Special Report from the Select Committee on the Daylight Saving Bill, August 24, 1909, xiii.
39. Hobsbawm, *Age of Empire,* 20.
40. BNA HO 45/11626, Summer Time—Notes of Deputation from the Early Closing Association, n.d., 11.
41. BNA HO 45/11940, Parliamentary Debates, House of Commons, Standing Committee C, Summer Time Bill, Official Report, March 24, 1925, 5, 6; BNA HO 45/19959, Report on the working of summer time in Ireland in 1917, n.d.
42. BNA HO 45/11940, Parliamentary Debates, House of Commons, Standing Committee C, Summer Time Bill, Official Report, March 24, 1925, 8, 18.
43. BNA HO 45/19959, "Daylight Saving Bill," to the Editor by "A working farmer," *Daily Chronicle,* February 19, 1918.
44. BNA HO 45/11626, letter, Ministry of Agriculture and Fisheries to Home Office, February 5, 1924; BNA HO 45/11077/411632, memorandum, Summer Time, Application to Scotland, n.d., received at Home Office, March 10, 1922.
45. BNA HO 45/11626, Summer Time—Notes of Deputation from the Early Closing Association, n.d., 2, 7; BNA HO 45/11626, newspaper clipping "Summer Time Demand," *Daily Chronicle,* November 1, 1923; newspaper clipping "Summer Time, Deputation to the Home Secretary," *The Times,* November 1, 1923.
46. BNA HO 45/10548/162178, Report and Special Report from the Select Committee on the Daylight Saving Bill, August 24, 1909, 29; BNA HO 45/11626, newspaper clipping "Summer Time Demand," *Daily Chronicle,* November 1, 1923.
47. Cross, *Quest for Time,* 80–86.
48. BNA HO 45/11893, letter, the National Amalgamated Union of Shop Assistants, Warehousemen, and Clerks to Home Office, May 18, 1916.
49. BNA HO 45/11893, letter, National Association of Goldsmiths to Home Office, May 16, 1916; BNA HO 45/11893, note, Shops—Question Put by Viscountess Astor, July 22, 1925; BNA HO 45/11893, letter, Home Office to National Association of Goldsmiths, June 6, 1916; newspaper clipping "New Clause Needed," *The Globe,* May 10, 1916; newspaper clipping "Willett Time Abuse," *The Daily Mail,* May 24, 1916; BNA HO 45/11893, letter, Home Office to Public Control Department, May 26, 1916.
50. See Pietro Basso, *Modern Times, Ancient Hours: Working Lives in the Twenty-First Century,* trans. Giacomo Donis (London: Verso Books, 2003); Jonathan

Crary, *24/7: Late Capitalism and the Ends of Sleep* (London: Verso Books, 2013).

51. BNA HO 45/10548/162178, Report and Special Report from the Select Committee on the Daylight Saving Bill, August 24, 1909, xi; unnamed position paper received at Home Office on March 24, 1914, 7.
52. George Darwin, "The Daylight Saving Bill: To the Editor of the Times," *Times of London,* July 8, 1908, 7.
53. "To the Editor of the Times," *Times of London,* July 11, 1908, 8.
54. HC Deb, May 8, 1916, vol. 82, c. 302–303.
55. BNA HO 45/10548/162178, unnamed position paper received at Home Office on March 24, 1914, 3; Report and Special Report from the Select Committee on the Daylight Saving Bill, August 24, 1909, xi; "The Daylight Saving Bill: To the Editor of the Times," *Times of London,* July 16, 1908, 16.
56. BArch R3001/7864, Nationalversammlung, 35th Session, April 11, 1919, 974; AP, VONC 20, Ministry of Public Works, Avance de l'heure légale pendant la belle saison, excerpt from Journal Officiel de la République Française, March 25, 1920, 201.
57. BNA HO 45/10548/162178, unnamed memorandum, received at Home Office, March 24, 1914, 8. On physiotemporal rhythms, see Zerubavel, *Hidden Rhythms,* 2; on present-day metronomic society, see Michael Young, *The Metronomic Society: Natural Rhythms and Human Timetables* (Cambridge, MA: Harvard University Press, 1988); and *The Rhythms of Society*, ed. Michael Young and Tom Schuller (London: Routledge, 1988).
58. BNA HO 45/10548/162178, Report and Special Report from the Select Committee on the Daylight Saving Bill, August 24, 1909, xii.
59. BArch R4702/207, Heinrich Grone, *Vorschlag zu einer Ersparung an künstlicher Beleuchtung durch verlängerte Ausnutzung des Tageslichtes* (Hamburg: Self-published, 1916).
60. BNA HO 45/10811/312364, leaflet, To Occupiers of Factories, May 18, 1916.
61. BArch R3001/7864, letter, Chancellor to Non-Prussian Minister Presidents and All Ministries, Berlin, March 1, 1917; BArch R3001/7864, letter, head of Imperial Office of Public Health to Secretary of State for the Interior, Berlin, October 24, 1918; BArch R4702/207, the Chancellor to Non-Prussian Minister Presidents and the Vice-Regent in Alsace Lorraine, Berlin, March 1, 1917.
62. BNA HO 45/10548/162178, Report and Special Report from the Select Committee on the Daylight Saving Bill, August 24, 1909, 93, 124.
63. Howse, *Greenwich Time,* 90; see Hannah Gay, "Clock Synchrony, Time Distribution and Electrical Timekeeping in Britain 1880–1925," *Past and Present* 181, no. 1 (2003): 107–140, here 120–121. A reference to Gay's wonderful piece was cut out inadvertently in editing my 2013 *American Historical Review* (hereafter *AHR*) article on time. See Vanessa Ogle, "Whose Time Is It? The Pluralization of Time and the Global Condition, 1870s–1940s," *AHR* 118, no. 5 (2013): 1376–1402.
64. "The Electric Telegraph," *The British Quarterly Review* 59 (January and April 1874): 438–469, here 466; "Correct Time: Post Office Methods of Distribution," *Times of London,* October 15, 1912, 4; RGO 7/252, letter, Standard Time

Company to Royal Astronomer, London, October 31, 1901; "The Greenwich Time Signal System," *Nature,* May 18, 1876, 50–52, and "The Greenwich Time Signal System: II," *Nature,* June 1, 1876, 110–113.

65. RGO 7/252, letter, Royal Astronomer to Chief Astronomer, Department of the Interior, Canada, September 17, 1901.
66. Gay, "Clock Synchrony," 123; RGO 7/252, letter, Royal Observatory to Standard Time Company, London, November 4, 1901; letter, The Standard Time Company to Royal Astronomer, London, February 17, 1902.
67. For the United States, Alexis McCrossen has termed the years from roughly the 1870s to the 1920s the "public clock era" due to the proliferation of clocks in the built environment. McCrossen, *Marking Modern Times;* British Telecom Archives (hereafter BTA), POST 30/1933, Time Signals, note, 10 a.m. Time Signal Direct from Greenwich, October 1909; Gay, "Clock Synchrony," 131, 135.
68. BTA POST 30/2536 Time Signals, internal note, Postmaster General, May 7, 1881.
69. BTA POST 30/2536, Time Signals, letter, Horace Darwin to General Post Office, December 31, 1881; BTA POST 30/2536, Greenwich Time Signals, internal note, Postmaster General, February 13, 1882, and ibid., General Post Office to Pearson, May 12, 1881.
70. RGO 7/252, newspaper clipping, "Greenwich Mean Time," *The Daily Graphic,* October 31, 1892. See also Gay, "Clock Synchrony," 118–120, and David Rooney, *Ruth Belville: The Greenwich Time Lady* (Greenwich: National Maritime Museum, 2008).
71. RGO 8/84, Royal Institution of Great Britain, Weekly Evening Meeting, speech by W. H. M. Christie, Astronomer Royal, "Universal Time," March 19, 1886, 2.
72. BTA POST 30/2536, newspaper clipping, E. J. D. Newitt, "Lying Clocks, To the Editor of the Times," *Times of London,* February 10, 1908, 12; "The Synchronization of Clocks," *Nature,* February 16, 1911, 516–517, here 516; "The Synchronization of Clocks," *Nature,* August 13, 1908, 353–354, here 353; BTA POST 30/1933, Daily Time Signals, letter, Hull Record Switch, Provision of Ringing Facilities, Engineer in Chief, Hull Record Switch, to General Post Office, December 9, 1909.
73. "Lying Clocks, To the Editor of the Times," *Times of London,* January 8, 1908, 9.
74. "Lying Clocks," *Times of London,* January 9, 1908, 7.
75. BTA POST 30/2536, newspaper clipping, E. J. D. Newitt, "Lying Clocks, To the Editor of the Times," *Times of London,* February 10, 1908, 12.
76. BTA POST 30/2536, newspaper clipping, Robert G. Orr, "Lying Clocks, To the Editor of the Times," *Times of London,* January 10, 1908, 12; "Lying Clocks," *Times of London,* November 9, 1908, 5; "The Synchronization of Clocks," *Times of London,* January 18, 1911, 9.
77. Quoted after Gay, "Clock Synchrony," 115. See Rooney, *Ruth Belville,* 74, 77.
78. Quoted after Howse, *Greenwich Time,* 114–115.

79. "Official Time," *Times of London,* July 9, 1924, 10; Gay, "Clock Synchrony," 139.
80. Thompson, "Time, Work Discipline, and Industrial Capitalism."
81. Paul Glennie and Nigel Thrift, "Reworking E. P. Thompson's 'Time, Work Discipline, and Industrial Capitalism,'" *Time and Society* 5, no. 3 (1996): 275–299, here 277, 284–285; as revisionist accounts, see Paul Glennie and Nigel Thrift, *Shaping the Day: A History of Timekeeping in England and Wales 1300–1800* (Oxford: Oxford University Press, 2009), who attempt to take apart Thompson on his own stomping ground. Thrift counts Landes *(Revolution in Time),* Harvey *(Condition of Postmodernity),* and Galison *(Einstein's Clocks)* among those who readily adopted Thompson's points and contributed to the establishment of a "Thompsonian" narrative. On historians of the United States who followed Thompson's notion of task time, see Michael O'Malley, "Time, Work, and Task Orientation: A Critique of American Historiography," *Time and Society* 1, no. 3 (1992): 341–358, here 343, 348. Others who have questioned Thompson's hypothesis are Gay, "Clock Synchrony," and Michael J. Sauter, "Clockwatchers and Stargazers: Time Discipline in Early Modern Berlin," *American Historical Review* 112, no. 3 (2007): 685–709. Hans Joachim Voth has questioned not so much Thompson's arguments about notations of time management but his source basis (largely literary). His own work is interested in the lengthening of the workday under industrial capitalism. See Hans Joachim Voth, "Time and Work in Eighteenth Century London," *Journal of Economic History* 58, no. 1 (1998): 29–58, here 30. See also Hans Joachim Voth, *Time and Work in England, 1750–1830* (Oxford: Oxford University Press, 2000). On clocks and time discipline in the slave South, see Mark M. Smith, *Mastered by the Clock: Time, Slavery, and Freedom in the American South* (Chapel Hill: University of North Carolina Press, 1997), who shows that clocks and clock discipline were easily combined with agrarian and plantation labor.
82. David Montgomery, *The Fall of the House of Labor: The Workplace, the State, and American Labor Activism, 1865–1925* (Cambridge: Cambridge University Press, 1987), esp. ch. 1 on "the manager's brain under the workman's cap."
83. Smithsonian Institution's National Museum for American History Archives, Warshaw Collection of Business Americana, Watch Works and Clock Works, collection no. 60, box 3, folder "Pneumatic," pamphlet advertising a pneumatic watchman clock.
84. Michel Foucault's work is emblematic for this set of arguments about disciplinary institutions; see Foucault, *Discipline and Punish,* 150–151, 156–162.
85. Thomas C. Smith, "Peasant Time and Factory Time in Japan," *Past and Present* 111, no. 1 (1986): 165–197.
86. On early modern time discipline in Berlin, see Sauter, "Clockwatchers and Stargazers."
87. Joseph Conrad, *The Secret Agent,* ed. John Lyon (Oxford: Oxford University Press, 2004).

88. Timothy Mitchell, *Rule of Experts: Egypt, Techno-Politics, Modernity* (Berkeley: University of California Press, 2002), 233, 14.

## 3. From National to Uniform Time around the Globe

1. Anthony Hopkins, "Back to the Future: From National History to Imperial History," *Past and Present* 164, no. 1 (1999): 198–243, here 238.
2. Giordano Nanni, *The Colonisation of Time: Ritual, Routine, and Resistance in the British Empire* (Manchester: Manchester University Press, 2012), 217; Daniel R. Headrick, *The Tools of Empire: Technology and European Imperialism in the Nineteenth Century* (New York: Oxford University Press, 1981), 10. The balance sheet for such technologies is overall more mixed, however, as telegraphs and railways could fall into the hands of colonial subjects especially in times of crisis.
3. The one source that to my knowledge lists all countries in the world with dates for the first adoption of mean times is Thomas G. Shanks, *The International Atlas: World Latitudes, Longitudes, and Time Changes* (San Diego: ACS Publications, 1985). Some of the information provided in Shanks is contradicted by archival evidence, hence other time indications might be inaccurate as well.
4. BNA CO 28/244, Government House, Barbados, to Minister of Colonies Joseph Chamberlain, December 22, 1897.
5. On Germany, see Andreas Daum, *Wissenschaftspopularisierung im 19. Jahrhundert: Bürgerliche Kultur, naturwissenschaftliche Bildung und die deutsche Öffentlichkeit, 1848–1914* (München: Oldenburg, 2002), esp. ch. 3; on Britain, see Bernard Lightman, *Victorian Popularizers of Science* (Chicago: University of Chicago Press, 2007), and later, Peter J. Bowler, *Science for All: The Popularization of Science in Early Twentieth-Century Britain* (Chicago: University of Chicago Press, 2009); on the United States, see Ronald C. Tobey, *The American Ideology of National Science, 1919–1930* (Pittsburgh: University of Pittsburgh Press, 1971).
6. BNA CO 28/244, Government House, Barbados, to Joseph Chamberlain, December 22, 1897.
7. BNA MT 9/652, copy of a minute by the Chief Surveyor, March 16, 1900.
8. BNA MT 9/652, Colonial Secretary for Singapore to Resident-General, Federated Malay States, Singapore, April 20, 1900; BNA MT 9/652, Colonial Secretary for Singapore to Board of Trade, Singapore, September 8, 1900; BNA CO 273/588/7, memorandum Daylight Savings Ordinance, December 31, 1932.
9. BArch R4701/207, newspaper clipping, Ostchinesischer Loyd, January 2, 1903; R901/37727, German Consulate Singapore to Foreign Office, Singapore, January 5, 1905.
10. RGO 7/146, Royal Observatory at Cape of Good Hope to Secretary of the Admiralty, Cape Town, February 2, 1892.
11. RGO 7/252, Colonial Office to Astronomer Royal, London, July 10, 1891; RGO 7/252, Colonial Office to Astronomer Royal, London, December 23, 1892; BArch R1001/6190 Ministry of Foreign Affairs to German Government

in Windhoek, Berlin, April 15, 1903; Tanzania National Archives (hereafter TNA) G8/12, German consulate in Mozambique to German government in Dar es Salaam, July 7, 1912.

12. BArch R1001/6190 High Commissioner South Africa, Johannesburg, September 1902; Imperial Postal Services, Swakopmund, November 5, 1902; Imperial Post Office, Windhoek, October 8, 1902; letter, January 26, 1907.
13. BArch R1001/6190, Foreign Office to German Government Windhoek, Berlin, April 15, 1903.
14. National Archives of Namibia (hereafter NAN) ZBU 81: A. II. vol. 1, Imperial Post Office to Imperial Railway Directorate, Swakopmund, November 5, 1902; BArch R1001/6190, Windhoek, February 1909.
15. BArch R1001/6190, Secretary of State at Colonial Office, Berlin, December 19, 1908.
16. NAN BWI 26: B. II. n, letter, Windhoek, February 16, 1909.
17. NAN BWI 26: B. II. n, Secretary of State at Colonial Office to Governor of German Southwest Africa in Windhoek, Berlin, March 12, 1909; BSW 72: B. II. m, letter, Swakopmund, November 26, 1912. Mayor of Swakopmund to District Office, November 25, 1912; District Office Lüderitzbucht to District Office Swakopmund, December 11, 1912.
18. BArch R1001/6190 Government of Togo, Lome, January 7, 1914; Berlin, February 21, 1914; Government of Togo, Lome, April 12, 1914.
19. BArch R1001/6190, Dar es Salaam, January 12, 1912; TNA G8/12, letter to Colonial Office, January 14, 1913.
20. BArch R1001/6190, East-African Railroad Company, Dar es Salaam, January 12, 1912; memorandum, March 19, 1912.
21. BArch R1001/6190, Dar es Salaam, April 11, 1913.
22. TNA G8/12, Dar es Salaam, n.d., April 1913; letter, Dar es Salaam, June 5, 1913; RGO 8/84, Colonial Office to Astronomer Royal, May 21, 1919; TNA G8/12, to Colonial Office, Dar es Salaam, January 14, 1913.
23. BNA CO 28/277, Government House Barbados to Colonial Office, October 3, 1911; Shanks, *International Atlas,* 80.
24. On the Mandate System, see Susan Pedersen, "Back to the League of Nations: Review Essay," *American Historical Review* 112, no. 4 (October 2007): 1091–1117, and Pedersen, "Getting Out of Iraq—in 1932: The League of Nations and the Road to Normative Statehood," *American Historical Review* 115, no. 4 (2010): 975–1000, and Pedersen's forthcoming book on the topic, *The Guardians: The League of Nations and the Crisis of Empire* (Oxford: Oxford University Press, 2015).
25. RGO 8/84, An Ordinance to Amend the Interpretation Ordinance, 1914, September 1, 1919; Colonial Office to Astronomer Royal, August 15, 1919; The Determination of Time Ordinance, Gold Coast, November 24, 1919; Colonial Office to Royal Observatory, February 7, 1933; Colonial Office to Royal Observatory, August 23, 1928, Colonial Office to Royal Observatory, January 14, 1926; Royal Observatory to Colonial Office, August 24, 1928; Colonial Office to Royal Observatory, July 18, 1936; Astronomer Royal to Colonial Office, July 21, 1936.

26. As one example of Argentine negotiations with scientific modernity, see Julia Rodriguez, *Civilizing Argentina: Science, Medicine, and the Modern State* (Chapel Hill: University of North Carolina Press, 2006).
27. Nicolás Besio Moreno and Estban Larco, "Husos horarios: hora legal en Argentina," *Anales de la sociedad científica Argentina* 70 (July 1910): 397–399, here 398; Richard H. Tucker, "The San Luis Observatory of the Carnegie Institution," *Publications of the Astronomical Society of the Pacific* 24, no. 140 (February 1912): 15–51.
28. Federico Villareal, "Informe del catedrático de la facultad de ciencias Dr. Don Federico Villareal, sobre la conveniencia de adoptar un meridiano Universal," *Anales universitarios de Perú publicados por el Doctor D. Francisco Rosas* 18 (1891) (Lima: Imprenta de F. Masias y Cia., 1891): 25–35, here 25.
29. Ibid., 27, 31–32.
30. Eulogio Delgado, "Hora oficial en el Perú (Standard Time)," *Boletín de la sociedad geográphica de Lima* 22 (1907): 39–41, here 39.
31. Fortunato L. Herrera, "El 'Standard Time' y la hora oficial para el Cuzco," *Revista de ciencias: Publicacion periodica redactada por professores de la facultad de ciencias y escuela de ingenerios* 13, no. 8 (August 1910): 174–178, here 175; and Fortunato L. Herrera, "El 'Standard Time' y la hora oficial para el Cuzco (concluye)," *Revista de ciencias: Publicacion periodica redactada por professores de la facultad de ciencias y escuela de ingenerios* 9, no. 9 (October 1910): 205–210, here 208.
32. Ronaldo Mourão and Rogério de Freitas, "Hora legal no Brasil e no mundo," *Revista do instituto histórico e geográfico Brasileiro* 434 (January–March 2007): 159–188, here 170; Antonio de Paula Freitas, "O meridiano inicial," *Boletim da sociedade de geographia do Rio de Janeiro* 1, no. 1 (1885): 161–175, here 171; Mônica Martins and Selma Junqueira, "A legalização da hora e a industrialização no Brasil," paper presented at the *XXI Jornadas de História Ecónomica,* Buenos Aires, September 2008, 6, 11.
33. Mourão and Freitas, "Hora legal," 171, 178–181; Martins and Junqueira, "Legalização da hora," 5; RGO 8/84, Republica Oriental del Uruguay to Royal Astronomer, Montevideo, April 26, 1920; Virgilio Raffinetti, *Descripción de los instrumentos astronómicos del observatorio de La Plata, seguida de una nota sobre los adelantos más recientes de la astrônoma* (La Plata: Talleres de Publicaciones del Museo, 1904), 197.
34. BArch R3001/7864, memorandum, Time-keeping at Sea, passed on to German authorities, January 26, 1920; "Standard Time at Sea," *The Geographical Journal* 51, no. 2 (February 1918): 97–100.
35. RGO 8/84, Bermuda, Time Zone Act, November 8, 1929; RGO 8/84, Colonial Office to Royal Astronomer, July 6, 1932.
36. BArch R1001/6190, undated memorandum, German-East-African Uniform Time.
37. BArch R1001/6190, Imperial High Court, Windhoek, May 9, 1906; order, Windhoek, September 18, 1906; Windhoek, September 10, 1906.
38. RGO 8/84, Colonial Office to Royal Observatory, November 18, 1932.

39. BNA CO 267/639, Government House Sierra Leone to Colonial Office, October 17, 1932; internal note, November 9, 1932; Colonial Office to Royal Observatory, March 19, 1938.
40. BNA CO 267/687, An Ordinance to Amend the Alteration of Time Ordinance, 1932, February 7, 1946.
41. BNA CO 533/380/8, Kenya Legislative Council, Report of Select Committee on Daylight Saving, n.d., 28.
42. BNA CO 533/380/8, Kenya Legislative Council, Motions, Daylight Saving, May 11, 1928, 72.
43. Ibid., 79.
44. Ibid., 74.
45. Ibid., 80, 75.
46. Ibid., 88, 96; RGO 8/84, undated Note to Mariners; see also Colonial Office to Astronomer Royal, February 18, 1930.
47. BNA DO 35/1123, Statutory Rules, 1942, No. 392, Regulation under the National Security Act 1939–1940, September 10, 1942; Statutory Rules, 1943, No. 241, September 29, 1943; newspaper clipping "Daylight Saving Dropped," no source, registered July 1, 1944; newspaper clipping "Australia to Drop Summer Time," *The Times of Australia,* August 28, 1944.
48. BNA DO 35/1123, High Commissioner's Office Pretoria to Dominion Office, Pretoria, September 5, 1943; FO 371/41524, Resident Minister Cairo to Foreign Office, February 14, 1944; FO 371/24320, Home Office to Foreign Office, January 11, 1940, on Cyprus; see also FO 371/46118, March 26, 1945; RGO 8/84, Colonial Office to Royal Observatory, June 20, 1940, on Malta.
49. Keletso Atkins, *The Moon Is Dead! Give Us Our Money! The Cultural Origins of an African Work Ethic, Natal, South Africa, 1843–1900* (Portsmouth, NH: Heinemann, 1993); see also Frederick Cooper, "Colonizing Time: Work Rhythms and Labor Conflict in Colonial Mombasa," in *Colonialism and Culture,* ed. Nicholas Dirks (Ann Arbor: University of Michigan Press, 1992), 209–245; Alamin Mazrui and Lupenga Mphande, "Time and Labor in Colonial Africa: The Case of Kenya and Malawi," in *Time in the Black Experience,* ed. Joseph K. Adjaye (Westport, CT: Greenwood Press, 1994), 97–119, here 98. On missionaries, colonial labor, and time, see Nanni, *Colonisation of Time,* 200.
50. Atkins, *The Moon Is Dead!,* 80, 81.
51. Nanni, *Colonisation of Time,* 3, 192. John L. Comaroff and Jean Comaroff, *Of Revelation and Revolution,* vol. 2: *The Dialectics of Modernity on a South African Frontier* (Chicago: University of Chicago Press, 1997), 300.
52. BNA FO 286/712, French Foreign Ministry to Foreign Office, April 26, 1919.
53. BNA FO 286/725, British High Commission Constantinople to Foreign Office, May 13, 1920.
54. BNA FO 371/110642, British Legation Seoul to Foreign Office, March 22, 1954, 1.
55. Ibid., 2.
56. Ibid.

57. BNA CO 323/815, Royal Observatory to Colonial Office, July 22, 1919; CO 323/904, Bureau International de l'Union Telegraphique to General Post Office, Berne, April 18, 1923; General Post Office to Colonial Office, April 27, 1923; undated memorandum, Heure légale in British colonies; DO 35/1123, Royal Observatory to Dominions Office, January 26, 1944.
58. John Darwin, *The Empire Project: The Rise and Fall of the British World-System, 1830–1970* (Cambridge: Cambridge University Press, 2009), esp. 3.
59. BNA DO 35/1123, Table Showing Standard Times Kept in the Dominions; RGO 8/84, The Robert Ramsay Organization to Royal Observatory, New York, April 9, 1930.
60. See Nils Gilman, *Mandarins of the Future: Modernization Theory in Cold War America* (Baltimore: Johns Hopkins University Press, 2003), and David Ekbladh, *The Great American Mission: Modernization and the Construction of an American World Order* (Princeton, NJ: Princeton University Press, 2010), as two examples of a ballooning literature on the history of development efforts in the so-called third world. On war and technological innovation, see the chapter on typewriters in Friedrich Kittler, *Gramophone, Film, Typewriter,* trans. Geoffrey Winthrop-Young and Michael Wutz (Stanford, CA: Stanford University Press, 1999).
61. RGO 8/84, Royal Observatory to Imperial Airways LTD, March 19, 1937.
62. Edgar Ansel Mowrer and Marthe Rajchman, *Global War: An Atlas of World Strategy* (New York: W. Morrow and Co., 1942); Leonard Oscar Packard, Bruce Overton, and Ben D. Wood, *Our Air-Age World: A Textbook in Global Geography* (New York: Macmillan Co., 1944); George T. Renner, *Global Geography* (New York: Thomas Y. Crowell, 1944); Grace Croyle Hankins, *Our Global World: A Brief Geography for the Air Age* (New York: Gregg, 1944); Leon Belilos, *From Global War to Global Peace* (Alexandria: Impr. du commerce, 1944); Henry Harley Arnold, *Global Mission* (New York: Harper, 1949); Herbert Rosinski, *India and the Global War* (Cambridge, MA: American Defense, Harvard Group, 1942).

## 4. A Battle of Colonial Times

1. Darwin, *Empire Project,* 10; Bright and Geyer, "Regimes of World Order," 217.
2. Sven Beckert, "Emancipation and Empire: Reconstructing the Worldwide Web of Cotton Production in the Age of the American Civil War," *American Historical Review* 109, no. 5 (2004): 1405–1438. On the Suez Canal, see Headrick, *Tentacles of Progress,* 26; Valeska Huber, *Channeling Mobilities: Migration and Globalisation in the Suez Canal Region and Beyond, 1869–1914* (Cambridge: Cambridge University Press, 2013). On Bombay's transformation, see Raj Chandavarkar, *The Origins of Industrial Capitalism in India: Business Strategies and the Working Classes in Bombay, 1900–1940* (Cambridge: Cambridge University Press, 2002), 23. On the history of Bombay generally, see Christine Dobbin, *Urban Leadership in Western India: Politics and Communities in Bombay City, 1840–1885* (London: Oxford University Press, 1972), es-

pecially on Bombay's merchant communities; Meera Kosambi, *Bombay in Transition: The Growth and Social Ecology of a Colonial City, 1880–1980* (Stockholm: Almqvist & Wiksell Intl., 1986); Teresa Albuquerque, *Urbs Prima in Indis: An Epoch in the History of Bombay, 1840–1865* (New Delhi: Promilla, 1985); Prashant Kidambi, *The Making of an Indian Metropolis: Colonial Government and Public Culture in Bombay, 1890–1920* (Aldershot, UK: Ashgate, 2007); and Gyan Prakash, *Mumbai Fables* (Princeton, NJ: Princeton University Press, 2010).

3. Chandavarkar, *Origins of Industrial Capitalism,* 29.
4. Kidambi, *Making of an Indian Metropolis,* 17.
5. "The Story of the Clock," *Bombay Gazette,* May 21, 1883, 3.
6. On the events in Bombay, see James Masselos's insightful article, "Bombay Time," in *Intersections: Socio-Cultural Trends in Maharastra,* ed. Meera Kosambi (Delhi: Orient Longman, 2000), 161–183, here 164.
7. British Library Asia, Pacific, and Africa Collections, India Office Records (hereafter BL APAC IOR), P/5664, memorandum, On Time in India: A Suggestion for Its Improvement. By R. D. Oldham, F. G. S., Superintendent, Geological Survey of India.
8. BL APAC IOR P/1781, Resolution, November 8, 1881; see also Masselos, "Bombay Time."
9. "'Standard' Time and Lord Lamington's First Serious Blunder," *Kaiser-i-Hind,* January 28, 1906, 12–14, here 12.
10. BL APAC IOR L/R/5/136, Report on Native Papers 1881, Bombay Presidency, *Indu Prakásh,* November 14, 1881; Masselos, "Bombay Time," 167.
11. "Bombay vs. Madras," *Bombay Gazette,* December 9, 1881, 2; *Bombay Gazette,* December 12, 1881, 3. See also Masselos, "Bombay Time," 169.
12. BL APAC IOR L/R/5/136, Report on Native Papers 1881, Bombay Presidency, *Bombay Samáchár,* 2. December 1881; *Times of India,* December 6, 1881, 2.
13. "Madras Time," *Times of India,* November 22, 1881, 5.
14. *Bombay Gazette,* March 23, 1882, 2.
15. "True Time and False Time," *Bombay Gazette,* May 17, 1882, 3.
16. "The University Clock," *Bombay Gazette,* April 7, 1882, 2; BL APAC IOR L/R/5/138, Report on Native Papers 1883, Bombay Presidency, *Jam-é-Jamshed,* April 11, 1883; Masselos, "Bombay Time," 172.
17. *Bombay Gazette,* May 12, 1883, 4; BL APAC IOR L/R/5/138, Report on Native Papers 1883, Bombay Presidency, *Indian Spectator,* April 29, 1883; *Indu Prakásh,* April 30, 1883; "The Rajabai Clock Tower," *Bombay Gazette,* April 19, 1883, 3; "Bombay vs. Official Time," *Times of India,* April 16, 1883, 5.
18. Masselos, "Bombay Time," 173.
19. BL APAC IOR P/5664, Proceedings of the Department of Revenue and Agriculture, 1899. Meteorology, Note on a proposal for an Indian Standard Time, para. 14.
20. BL APAC IOR P/5664, Proceedings of the Department of Revenue and Agriculture, 1899. Meteorology, letter, John Milne, Secretary, British Association Committee, to Under-Secretary of State for India, October 9, 1897; John Milne, "Civil

Time; Or Tables Showing the Difference in Time between that Used in Various Parts of the World and Greenwich Mean Time," *The Geographical Journal* 13, no. 2 (February 1899): 173–194; BL APAC IOR P/5437, Proceedings of the Department of Revenue and Agriculture, 1898. Meteorology, Despatch, Secretary of State for India, November 10, 1898; letter, Royal Scottish Geographical Society, Edinburgh, June 7, 1898.

21. BL APAC IOR P/5664, Proceedings of the Department of Revenue and Agriculture, 1899. Meteorology, Royal Scottish Geographical Society, memorandum, Notes on Standard Time.
22. BL APAC IOR P/5664, Proceedings of the Department of Revenue and Agriculture, 1899; Meteorology, memorandum, On Time in India: A Suggestion for Its Improvement. By R. D. Oldham, Superintendent, Geological Survey of India.
23. "Time in India," *The Pioneer,* June 18, 1899, 7.
24. BL APAC IOR P/5664, Proceedings of the Department of Revenue and Agriculture, 1899; Meteorology, memorandum, On Time in India: A Suggestion for Its Improvement. By R. D. Oldham, Superintendent, Geological Survey of India.
25. BL APAC IOR P/5664, Notes from the Council of the Royal Geographical Society on the Subject of a "Standard Time" for all India.
26. BL APAC IOR P/5664, GOI, Proceedings of the Department of Revenue and Agriculture, 1899; Meteorology, memorandum, On Time in India: A Suggestion for Its Improvement. By R. D. Oldham, Superintendent, Geological Survey of India.
27. BL APAC IOR P/5664, Proceedings of the Department of Revenue and Agriculture, 1899; Meteorology, letter, George F. Hamilton, Secretary of State for India, Simla, August 10, 1899. See also Hamilton to Asiatic Society, September 1, 1899, with similar reasons stated.
28. BL APAC IOR P/6828, letter, John Eliot, Meteorological Reporter to Secretary to the Government of India, Simla, April 28, 1903.
29. BL APAC IOR P/6828, letter, J. Wilson, Secretary to the Government of India, Simla, July 13, 1904.
30. BL APAC IOR P/5664, GOI, Proceedings of the Department of Revenue and Agriculture, 1899. Meteorology, Note on a proposal for an Indian Standard Time, para. 5, 11.
31. See, for example, a list of twenty-seven firms, provided by the Karachi Chamber: *Report of the Karachi Chamber of Commerce for the Year Ending 31st December 1905, Presented at the Annual General Meeting Held on the 17th April 1906* (Karachi, 1906), 326.
32. BL APAC IOR P/7073, GOI, Proceedings of the Department of Revenue and Agriculture, 1905. Meteorology, letter, J. N. Atkinson, Acting Secretary to the Government of Madras to Secretary to the Government of India, Ootacamund, September 30, 1904.
33. BL APAC IOR P/7073, Proceedings of the Department of Revenue and Agriculture, 1905. Meteorology, letter, Secretary to the Government of Bombay, General Department, to Secretary to the Government of India, Bombay Castle, December 29, 1904; see also "Indian Standard Time," *Amrita Bazar Patrika,* February 3, 1905, 4.
34. "Standard Time," *Amrita Bazar Patrika,* May 27, 1905, 6.

35. Masselos, "Bombay Time," 176–177.
36. BL APAC IOR P/7073, letter, Secretary of State for India, April 27, 1905. See also ibid., press communiqué, June 1, 1905; Masselos, "Bombay Time," 175.
37. On municipal reform and the possibilities of nationalism in the city, see Prashant Kidambi, "Nationalism and the City in Colonial India: Bombay, c. 1890–1940," *Journal of Urban History* 38, no. 5 (2012): 950–967, here 953.
38. *Report of the Bombay Chamber of Commerce for the Year 1905: Presented to the Annual General Meeting Held on the 1st March 1905* (Bombay, 1906), printed documents, 354–355; Director, Government Observatory, to Secretary to the Government, General Department, Colaba, September 10, 1904; BL APAC IOR P/6828, GOI, Proceedings of the Department of Revenue and Agriculture, 1904; Meteorology, letter, John Eliot, Meteorological Reporter to the Government of India to Secretary to the Government of India, Simla, April 28, 1903.
39. BL APAC IOR P/7073, GOI, Proceedings of the Department of Revenue and Agriculture, 1905. Meteorology, J. Sladen, Secretary to the Government of Bombay, to Secretary to the Government of India, October 17, 1905.
40. BL APAC IOR L/R/5/160, Report on Native Papers 1905, Bombay Presidency, *Jám-e-Jamshed,* September 2, 1905; "Standard Time or Zone Time?," *Bombay Gazette,* July 27, 1905, 3; "Standard Time in India," *Amrita Bazar Patrika,* September 1, 1905, 3.
41. BL APAC IOR P/7187, General Department Proceedings for the Year 1905, letter, President, Municipal Corporation, October 16, 1905; "Bombay Corporation. Adoption of the 'Standard Time,'" *Bombay Gazette,* January 23, 1906, 6; BL APAC IOR L/R/5/161, Report on Native Papers 1906, Bombay Presidency, *Oriental Review,* January 24, 1906.
42. "'Standard' Time and Lord Lamington's First Serious Blunder," 14.
43. On institutional Indian "nationalism" in Bombay, see Gordon Johnson, *Provincial Politics and Indian Nationalism: Bombay and the Indian National Congress 1880–1915* (Cambridge: Cambridge University Press, 1973). On Mehta and his involvement with Bombay politics, see James Masselos, *Towards Nationalism: Group Affiliations and the Politics of Public Associations in Nineteenth Century Western India* (Bombay: Popular Prakashan, 1974), esp. 240.
44. J. R. B. Jeejeebhoy, *Some Unpublished and Later Speeches and Writings of the Hon. Sir Pherozeshah Mehta* (Bombay: Commercial Press, 1918): 177–178. On the Chamber of Commerce and similar associations, see Kidambi, *Making of an Indian Metropolis,* 188.
45. "Return to Bombay Time," *The Tribune,* April 25, 1906, 3; "Standard Time in Bombay," *Amrita Bazar Patrika,* April 25, 1906, 3; BL APAC IOR L/R/5/161, Report on Native Papers 1906, Bombay Presidency, *Indu Prakásh,* April 25, 1906. See also V. S. Srinivasa Sastri, *Life and Times of Sir Pherozeshah Mehta* (Bombay: Bharatiya Vidya Bhavan, 1975), 102.
46. BL APAC IOR L/R/5/160, Report on Native Papers 1905, Bombay Presidency, *Kaiser-i-Hind,* June 11, 1905.
47. BL APAC IOR L/R/5/160, Report on Native Papers 1905, Bombay Presidency, *Kaiser-i-Hind,* August 13, 1905; "Standard Time," *Bombay Gazette,* August 2, 1905, 4; *Gujaráti,* June 18, 1905; BL APAC IOR P/7073, GOI, Proceedings of

the Department of Revenue and Agriculture, 1905, Meteorology, Resolution passed at the General Meeting of the Members of the Bombay Native Piece Goods Merchants' Association, September 17, 1905; BL APAC IOR L/R/5/160, Report on Native Papers 1905, Bombay Presidency, *Oriental Review,* September 20, 1905; BL APAC IOR P/7073, Proceedings of the Department of Revenue and Agriculture, 1905; Meteorology, letter, Kukhamsee Nappoo, Chairman of the Grain Merchants' Association, to Chief Secretary to the Government of Bombay, October 13, 1905.

48. BL APAC IOR L/R/5/136, Report on Native Papers 1881, Bombay Presidency, *Indu Prakásh,* November 14, 1881; "Madras Time in Bombay," *Bombay Gazette,* November 21, 1881, 3; *Jám-e-Jamshed,* November 24, 1881.
49. BL APAC IOR P/7073, GOI, Proceedings of the Department of Revenue and Agriculture, 1905; Meteorology, letter, W. Parsons, Secretary, Bengal Chamber of Commerce, Calcutta, to Secretary to the Government of Bengal, October 13, 1904.
50. BL APAC IOR L/R/5/160, Report on Native Papers 1905, Bombay Presidency, *Gujaráti,* June 18, 1905.
51. BL APAC IOR L/R/5/160, Report on Native Papers 1905, Bombay Presidency, *The Phoenix,* July 5, 1905.
52. Ibid.
53. BL APAC IOR P/7457, GB, General Department Proceedings for the Year 1907, letter, Secretary to the Government to Ahmedbhoy Habibbhoy, January 19, 1906; "Standard Time in Bombay," *Amrita Bazar Patrika,* January 22, 1906, 3.
54. "Bombay 'Slaves' and Standard Time," *Bombay Gazette,* January 15, 1906, 7; "Turbulent Mill Hands," *Times of India,* January 6, 1906, 6; "Standard Time in Bombay," *Amrita Bazar Patrika,* January 22, 1906, 3.
55. BL APAC IOR L/R/5/161, Report on Native Papers 1906, Bombay Presidency, *Jám-e-Jamshed,* January 5 and 6, 1905. See also "Standard Time. Dissatisfaction among Bombay Mill Hands. A Strike Averted," *Bombay Gazette,* January 6, 1906, 3; "Standard Time in Bombay: Strike of Mill-Hands," *Amrita Bazar Patrika,* January 8, 1906.
56. "Disapproval of the 'Stupid' Time by the Towns People," *Kaiser-i-Hind,* February 25, 1906, 14; BL APAC IOR L/R/161, Report on Native Papers 1906, *Oriental Review,* February 21, 1906. See also *Bombay Gazette,* "Meeting at Madhav Baug. Protest against Standard Time," February 22, 1906, 4.
57. BL APAC IOR L/R/5/160, Report on Native Papers 1905, Bombay Presidency, *Kaiser-i-Hind,* September 3, 1905; "'Standard' Time and Lord Lamington's First Serious Blunder," 13.
58. John Darwin, *Unfinished Empire: The Global Expansion of Britain* (London: Allen Lane, 2012), 212.
59. "'Standard' Time and Lord Lamington's First Serious Blunder," 13.
60. BL APAC IOR P/7073, GOI, Proceedings of the Department of Revenue and Agriculture, 1905. Meteorology, letter, Kukhamsee Nappoo, Chairman of the Grain Merchants' Association, to Chief Secretary of the Government of Bombay, October 13, 1905.

61. BL APAC IOR P/7073, GOI, Proceedings of the Department of Revenue and Agriculture, 1905; Meteorology, Resolution passed at the General Meeting of the Members of the Bombay Native Piece Goods Merchants' Association, September 17, 1905; BL APAC IOR L/R/5/160, Report on Native Papers 1905, Bombay Presidency, *Oriental Review,* September 20, 1905; *Gujaráti,* September 24, 1905; *Kaiser-i-Hind,* August 13, 1905.
62. Sumit Sarkar, *The Swadeshi Movement in Bengal, 1903–1908* (Delhi: People's Publishing House, 1973), esp. chs. 2 and 9.
63. BL APAC IOR L/R/5/160, Report on Native Papers 1905, Bombay Presidency, *Kaiser-i-Hind,* December 24, 1905.
64. BL APAC IOR L/R/5/161, Report on Native Papers 1906, Bombay Presidency, *Oriental Review,* January 10, 1906; "'Standard' Time and Lord Lamington's First Serious Blunder," 12.
65. BL APAC IOR L/R/5/160, Report on Native Papers 1905, Bombay Presidency, *Kaiser-i-Hind,* June 11, 1905; *Gujaráti,* December 17, 1905.
66. "'Standard' Time and Lord Lamington's First Serious Blunder," here 12.
67. "Meeting at Madhav," *Times of India,* February 22, 1906, p. 3.
68. "The Senseless Standard Time Once More!," *Kaiser-i-Hind,* June 28, 1908, 13; see also Masselos, "Bombay Time," 180.
69. "Bombay Time or Standard: Corporate Debate," *Times of India,* June 24, 1927, 10.
70. "Bombay Time: Difficulties of Observance," *Times of India,* June 26, 1928, 7; "Bombay Time Retained: City Corporation Advocates of Standard Time in Minority," *Times of India,* September 27, 1929, 11.
71. "Bombay Time to Stay: Corporation Decision," *Times of India,* June 30, 1939, 6; "New Time Not for Corporation: Proposal Rejected," *Times of India,* September 3, 1942, 5; "Prolonged Debate on Municipal Time. City Corporation Inquiry into Property Tax Collection," *Times of India,* June 23, 1939, 17.
72. "Battle of Clocks in Corporation: Bombay Time to Stay," *Times of India,* August 23, 1935, 17; "Standard Time: To the Editor of the 'Times of India,'" *Times of India,* May 11, 1927, 15; "Bombay Time," *Times of India,* June 25, 1927, 10; "Bombay Municipality Adopts Standard Time: 44-Year-Old Battle of Clocks Ends," *Times of India,* March 15, 1950, 5.
73. *Times of India,* July 19, 1919, 10.
74. "Prolonged Debate on Municipal Time. City Corporation Inquiry into Property Tax Collection," *Times of India,* June 23, 1939, 17; Darwin, *Unfinished Empire,* 204.
75. On the nested nature of various time scales and planes that eventually interacted to form a notion of Indian national space and time, see Goswami, *Producing India,* chs. 5 and 6. On early nationalists' datings of national histories see also James Masselos, "Time and Nation," in *Thinking Social Science in India: Essays in Honour of Alice Thorner,* ed. Sujata Patel, Jasodhara Bagchi, and Krishna Raj (New Delhi: Sage Publications, 2002), 343–354, esp. 344.

## 5. Comparing Time Management

1. On cities, see Osterhammel, *Transformation,* ch. VI.
2. Greater Syria comprises present-day Lebanon, Syria, Israel/Palestine, Jordan, and parts of Iraq. It is used here as a geographical term, not in the way it was politicized as part of Arab nationalism in the twentieth century.
3. Jens Hanssen, *Fin de siècle Beirut: The Making of an Ottoman Provincial Capital* (Oxford: Oxford University Press, 2003), 27, 28. Leila Fawaz, *Merchants and Migrants in Nineteenth-Century Beirut* (Cambridge, MA: Harvard University Press, 1983), 29, gives a number of 6,000 at the beginning of the century and 120,000 at its closure; see also Fawaz, *Merchants and Migrants,* 63.
4. Hanssen, *Fin de siècle,* 33.
5. On the events of 1860 and their legacy, see Ussama Makdisi, *The Culture of Sectarianism: Community, History, and Violence in Nineteenth-Century Ottoman Lebanon* (Berkeley: University of California Press, 2000); Fawaz, *Merchants and Migrants,* 23–24; Hanssen, *Fin de siècle,* 35, 50–51, 164; see also Davide Rodogno, *Against Massacre: Humanitarian Interventions in the Ottoman Empire, 1815–1914: The Emergence of a European Concept and International Practice* (Princeton, NJ: Princeton University Press, 2012).
6. Fawaz, *Merchants and Migrants,* 60, 70; Hanssen, *Fin de siècle,* 39, 101.
7. Fawaz, *Merchants and Migrants,* 15–23. During the nineteenth century, the number of Shi'a residing in Beirut was extremely small. Roughly two-thirds of the city's population were Christian, one-third Muslim. Among the Christians, the Greek Orthodox constituted the largest group, followed closely by Maronites. The remaining Christians were Greek Catholic, Roman, Syrian, Armenian Catholic, Armenian Orthodox, and Protestant. Jews fluctuated between 1 and 3 percent. Fawaz, *Merchants and Migrants,* 51. The term "poly-rhythmic" is Henri Lefebvre's: see Lefebvre and Catherine Régulier, "Attempt at the Rhythmanalysis of Mediterranean Cities," in *Rhythmanalysis: Space, Time, and Everyday Life,* ed. Lefebvre (New York: Continuum, 2004), 85–100, here 100.
8. On the use of "Frankish" and "Turkish" time in the Ottoman Empire, see Avner Wishnitzer, "'Our Time:' On the Durability of the Alaturka Hour System in the Late Ottoman Empire," *International Journal of Turkish Studies* 16, nos. 1 and 2 (2010): 47–69, here 48. See also Barbara Stowasser, *The Day Begins at Sunset: Perceptions of Time in the Islamic World* (London: I. B. Tauris, 2014), 142.
9. Henry Jessup, *Fifty-Three Years in Syria* (Reading, UK: Garnet, 2002 [1910]), 331; Hanssen, *Fin de siècle,* 245.
10. Marwa Elshakry, *Reading Darwin in Arabic 1860–1950* (Chicago: University of Chicago Press, 2013), 53. The Jesuits opened an observatory in the Beqa'a Valley in 1907, but it specialized in meteorology and did not carry out astronomical timekeeping.
11. Jessup, *Fifty-Three Years,* 331.
12. Daniel Bliss, *Letters from a New Campus Written to His Wife Abby and Their Four Children During Their Visit to Amherst, Massachusetts, 1873–1874,* col-

lected and annotated by Douglas and Belle Dorman Rugh and Alfred H. Howell (Beirut: American University of Beirut, 1994), 59; American University of Beirut, Archives and Special Collections, Jafet Library (hereafter AUB), Buildings and Grounds: Individual Buildings, College Hall, AA: 2.5.3.3.2.3, box 5: General; Tower and Bell; Explosion, 1936–; file 3: Tower and Bell; excerpt and copies from "Prof. M. Jurdak Corrects F. B. Feature on College Hall Tower," AUB Faculty Bulletin vol. 6, January 26, 1963.

13. AUB, Buildings and Grounds, Individual Buildings, box 16: Jessup Hall; Lee Observatory; Marquand House; Medical Center, 1874; file 3: Lee Observatory, 1874–; excerpts from Najwa Shaheen Haffar, "The Observatory and the Stars," article taken from AUB's magazine *al-Kulliyya,* Autumn/Winter 1988, 11. See AUB, Buildings and Grounds, Individual Buildings, College Hall, AA: 2.5.3.3.2.3, box 5: General; Tower and Bell; Explosion, 1936–; file 3: Tower and Bell; draft manuscript for article later published in AUB's official newspaper, *Outlook.*
14. Selim Deringil, *The Well-Protected Domains: Ideology and the Legitimation of Power in the Late Ottoman Empire, 1876–1909* (London: I. B. Tauris, 1998).
15. On clock towers in Anatolia, see Mehmet Bengü Uluengin, "Secularizing Anatolia Tick by Tick: Clock Towers in the Ottoman Empire and the Turkish Republic," *International Journal of Middle East Studies* 42 (2010): 17–36. See also Klaus Kreiser, "Ottoman Clock Towers: A Preliminary Survey and Some General Remarks on Construction Dates, Sponsors, Locations, and Functions," in *Essays in Honor of Ekmeleddin Ihsanoglu,* vol. 1, ed. Mustafa Kacar and Zeynep Durukal (Istanbul: Research Centre for Islamic History, Art, and Culture, 2006), 545–547. The Sultan of Zanzibar built a clock tower in Zanzibar in 1879: see Roman Loimeier, *Eine Zeitlandschaft in der Globalisierung: Das islamische Sansibar im 19. und 20. Jahrhundert* (Bielefeld: Transcript, 2012), 90. On the symbolic and often very practical conflicts over church bells in France, see Alain Corbin, *Village Bells: Sound and Meaning in the Nineteenth-Century French Countryside,* trans. Martin Thom (New York: Columbia University Press, 1998). On the role of church bells in conflicts the Bolsheviks fought over collectivization in interwar Russia, see Richard L. Hernandez, "Sacred Sound and Sacred Substance: Church Bells and the Auditory Culture of Russian Villages during the Bolshevik Velikii Perelom," *American Historical Review* 109, no. 5 (December 2004): 1475–1504.
16. Quoted after Hanssen, *Fin de siècle,* 243. See also Jens Hanssen, "'Your Beirut Is on My Desk': Ottomanizing Beirut under Sultan Abdülhamid II (1876–1909)," in *Projecting Beirut: Episodes in the Construction and Reconstruction of a Modern City,* ed. Peter G. Rowe and Hashim Sarkis (Munich: Prestel, 1998), 41–67.
17. Academie Libanaise des Beaux-Arts ALBA, collection of letters and documents pertaining to Yussuf Aftimus, CD-ROM, letter, Aftimus to Edward Prest, November 26, 1937. See also Hanssen, *Fin de siècle,* 236–237.
18. Daniel Stolz, "The Lighthouse and the Observatory: Islam, Authority, and Cultures of Astronomy in Late Ottoman Egypt" (PhD diss., Princeton University, 2013), 299.

19. Wishnitzer, "Our Time," 53–57. Ahmet Hamdi Tanpinar's wonderful novel *The Time Regulation Institute* is a fictional account of the new Turkish state's attempt to "westernize" timekeeping after the end of the Ottoman Empire. See Tanpinar, *The Time Regulation Institute*, trans. Maureen Freely and Alexander Dawe (New York: Penguin, 2013). On temporal pluralism in prewar Istanbul, see also François Georgeon, "Temps de la réforme, réforme du temps: Les avatars de l'heure et du calendrier à la fin de l'Empire Ottoman," in *Les Ottomans et le temps,* ed. François Georgeon and Frédéric Hitzel (Leiden: Brill, 2012), 241–279.
20. Nikula Shahin (as transliterated by AUB's archive), an astronomer at AUB's Lee Observatory in the 1950s, mentioned in passing in a radio-broadcasted lecture that GMT was introduced to Beirut in 1917. See AUB, Nikula Jurjus Shahin Collection: AUB Faculty 1918–1966, box 1, file 5: Hiwar ʿilmi li-l-idhaʿa, 1.
21. As one such almanac, see *Taqwim al-Bashir 1928* (Beirut: Matbaʿa al-yasuʿiyin, 1927), 96; and *Taqwim al-Bashir 1931* (Beirut: Matbaʿa al-yasuʿiyin, 1930), 42, 44.
22. Ernest Weakley, *Report upon the Conditions and Prospects of British Trade in Syria* (London: Darling and Son, 1911), 163. On earlier imports of luxury clocks into the Ottoman Empire, see Ian White, *English Clocks for the Eastern Markets: English Clockmakers Trading in China and the Ottoman Empire 1580–1815* (Sussex: The Antiquarian Horological Society, 2012). Annual commercial statistics show varied patterns of clock and watch imports to the region, although the numbers might not be reliable. In 1902, 12,500 kilograms of clocks and 9,100 kilograms of watches are said to have been imported through the port of Beirut: Rapport commercial, Année 1902. In 1904, those numbers reached 17,000 kilograms (clocks) and 9,100 (watches): see Rapport commercial du consulat general de Beyrouth pour l'année 1904. In 1905, 2,100 kilograms of ordinary and 800 kilograms of fine watches were registered: see Turquie d'Asie, Mouvement commercial et maritime de Beyrouth en 1905, all in Archives du ministère des affaires étrangères, Nouvelle Série, Turquie, No. 479. See also Archives de la compagnie du port, des quais et entrepôts de Beyrouth, P8/H/5/C/2, Statistique générale 1912. In 1912, 17,400 kilograms of clocks and only 200 kilograms of ordinary watches were imported. For 1921, the numbers were 18,000 kilograms of clocks, 150 kilograms of ordinary watches, and 50 kilograms of "fine" watches; P8/H/5/D/1 Rapport annuels 1921.
23. Weakley, *Report,* 164.
24. ʿAbd al-Basit al-Unsi, *Dalil Bayrut wa Taqwim al-Iqbal li-Sanat 1327* (Beirut: al-Iqbal, 1910/1011), 147; Hanssen, *Fin de siècle,* 217.
25. "Iʿlan," *Lisan al-Hal,* December 31, 1897, 1; Muhammad Saʿid al-Qasimi, *Qamus al-Sinaʿa al-Shamiyya,* al-Juzʾ al-ʾAwwal (Paris, 1960), 174.
26. "Mukhtaraʿ al-Saʿat," *al-Hilal* 4, no. 15 (May 15, 1896): 700–702, and "Mukhtaraʿ al-Saʿat," *al-Hilal* 4, no. 22 (July 15, 1896): 857–858; *Lisan al-Hal,* August 19, 1898, 3; "Dabt al-Saʿat," *al-Jinan,* no. 8 (April 15, 1878): 252; "Saʿa Qadima," *Lisan al-Hal,* September 9, 1898, 4.
27. See "Saʿa Ghariba," *Thamarat al-Funun,* May 15, 1893, 4–5.

28. On the short story "Sa'at al-Kuku," see Jan Na'um Tannus, *Surat al-Gharb Fi al-'Adab al-'Arabi al-Mu'asir* (Beirut: Dar al-Manhal al-Lubnani, 2009), part I. Nu'ayma's name is transliterated in various ways; spellings such as Naimy are found occasionally.
29. Mikha'il Nu'ayma, "Sa'at al-Kuku," in Mikha'il Nu'ayma, *Kan ma Kan* (Beirut: Matba'a Sadir, 1968), 7–38, here 25–26.
30. On the clock story, see Akram Khater, "A History of Time in Mount Lebanon, 1860–1914," *Chronos: Revue d'histoire de l'Université de Balamand* 2 (1999): 131–155, here 151.
31. Joseph Harfouch, *Le Drogman Arabe ou guide pratique de l'Arabe parlé en charactères figurés pour la Syrie, la Palestine et l'Egypte* (Beirut: al-Matba'a al-Kathulikiyya, 1894), 205, 208, 243.
32. "Al-Hisaban al-Gharbi wa-l-Sharqi," *al-Hilal* 4, no. 16 (April 15, 1896): 618–620; Yussuf Ibrahim Khayr Allah, "Al-Hisaban al-Sharqi wa-l-Gharbi," *al-Hilal* 14, no. 2 (April 1, 1905): 422–423; Elias Estephan, "Al-Hisaban al-Sharqi wa-l-Gharbi wa-l-Farq Baynahuma," *al-Hilal* 2, no. 3 (October 1, 1893): 69–75; see also Ahmad Mukhtar Pasha, "Islah al-Taqwim," *al-Muqtataf* 14, no. 7 (April 1, 1890): 488–490; see Edward Van Dyck, "Al-Taqwim: Fi Tatbiq Mawaqit al-Bashar 'ala Dawran al-Shams wa-l-Qamr," *al-Muqtataf* 14, no. 10 (July 1, 1890): 660–667; Edward Van Dyck, "Al-Taqwim: Fi Tatbiq Mawaqit al-Bashar 'ala Dawran al-Shams wa-l-Qamr," *al-Muqtataf* 14, no. 11 (August 1, 1890), 735–743; Edward Van Dyck, "Al-Taqwim: Fi Mabadi' Ihtisab al-Sinnin," *al-Muqtataf* 14, no. 12 (September 1, 1890): 810–815. Van Dyck was a consular clerk of the United States in Cairo.
33. Jurji Zaydan, "Mahmud Pasha al-Falaki, al-'Alim al-Riadi al-Falaki al-Masri," in Jurji Zaydan, *Tarajim Mashahir al-Sharq fi al-Qarn al-Tasi' 'Ashar,* vol. 2 (Beirut: Dar Maktabat al-Hayat, 1967), 148–151; Mahmoud Effendi, *Memoire sur le calendrier arabe avant l'Islamisme, et sur la naissance et l'age du prophete Mohammad* (Paris: Imprimerie Impériale, 1858); Mahmud Pasha also authored a book on ancient and contemporary Egyptian weights and measures. On Mahmud Pasha, see Stolz, *Lighthouse,* ch. 2.
34. Ghazi Ahmed Moukhtar Pacha, *La Reforme du Calendrier* (Leiden: Brill, 1893). The book was published first in Ottoman Turkish and subsequently translated into Arabic and French.
35. "Shuruq al-Shams," *Thamarat al-Funun,* January 11, 1892, 2.
36. Ibid., and "Jawab al-Bashir," *Thamarat al-Funun,* January 18, 1892, 2.
37. "Shuruq al-Shams," *Thamarat al-Funun,* January 11, 1892, 2; " 'Idhah Mas'ala," *al-Bashir,* January 6, 1892, 3; the Arabic original says "mukhālif li-l-maḥsūs"; " 'Idhah Mas'ala," *al-Bashir,* January 6, 1892, 3.
38. "Jawab al-Bashir," *Thamarat al-Funun,* January 18, 1892, 2.
39. See Şükrü Hanioglu, *A Brief History of the Late Ottoman Empire* (Princeton, NJ: Princeton University Press, 2008), ch. 4; Fawaz, *Merchants and Migrants,* 107.
40. Juan Cole, *Colonialism and Revolution in the Middle East: Social and Cultural Origins of Egypt's 'Urabi movement* (Princeton, NJ: Princeton University Press,

1993); Alexander Schölch, *Egypt for the Egyptians! The Socio-Political Crisis in Egypt, 1878–1882* (London: Ithaca Press, 1981).

41. Attitudes toward the British in Egypt itself were more varied. Few voices outright rejected the British Protectorate and many openly embraced foreign rule and even defended it on the grounds that Britain was forced to secure the Suez Canal in this way. Elshakry, *Reading Darwin,* 95.
42. Ilham Khuri-Makdisi, *The Eastern Mediterranean and the Making of Global Radicalism, 1860–1914* (Berkeley: University of California Press, 2010). On the early origins of the *Nahḍa,* see in particular Abdulrazzak Patel, *The Arab Nahḍa: The Making of the Intellectual and Humanist Movement* (Edinburgh: Edinburgh University Press, 2013).
43. Elshakry, *Reading Darwin,* 47; on Butrus al-Bustani, see Butrus Abou Manneh, "The Christians Between Ottomanism and Syrian Nationalism: The Ideas of Butrus al-Bustani," *International Journal of Middle East Studies,* 11 (1980): 287–304; see also Zaydan, *Tarajim,* 35–44; Ami Ayalon, *The Press in the Arab Middle East: A History* (Oxford: Oxford University Press, 1995), 35.
44. Ayalon, *Press,* 34; on Khalil Sarkis, see Ami Ayalon, "Private Publishing in the Nahda," *International Journal of Middle East Studies* 40, no. 4 (2008): 561–577, here 565–566.
45. Elshakry, *Reading Darwin,* 25, 72; Ayalon, *Press,* 53; see Anne-Laure Dupont, *Gurgi Zaydan (1861–1914): Ecrivain réformiste et témoin de la renaissance Arabe* (Damascus: IFPO, 2006), esp. ch. 10; Ayalon, *Press,* 53.
46. I borrow this argument from Elshakry, *Reading Darwin,* 9. On Freemasonry in the Middle East, see Johann Büssow, *Hamidian Palestine: Politics and Society in the District of Jerusalem, 1872–1908* (Leiden: Brill, 2011); Elshakry, *Reading Darwin,* 22.
47. Archives de la Compagnie de Jésus à Beyrouth, fol. 12 A6, "Al-Bashir." Excerpts from "Le Jésuite en Syrie," here 27: in certain villages, only one subscriber existed—the head of the school, for instance. The entire local population borrowed the newspaper through one subscriber. See also generally Elshakry, *Reading Darwin,* 22.
48. The most well-known caricature and criticism of a Westernizing Arab is probably "ʿArabi Tafarnaj" by ʿAbd Allah al-Nadim, Egyptian political activist and journalist. See Samah Selim, *The Novel and the Rural Imaginary in Egypt* (New York: RoutledgeCurzon, 2004), 50; Jubran Massuh, "Kayfa naqtul al-waqt," *Lisan al-Hal,* August 14, 1909, 3.
49. Massuh, "Kayfa naqtul al-waqt," 3.
50. Ibid.
51. "Awqat al-Taʿam," *al-Muqtabas* 1 (1906): 31–32, here 31.
52. Ibid., 32.
53. Ibid.
54. Salim Diyab, "Mulaqat al-Zaman," *al-Jinan,* no. 19 (October 1, 1875): 665–669, here 669.
55. Seema Alavi, "'Fugitive Mullahs and Outlawed Fanatics'": Indian Muslims in Nineteenth-Century Trans-Asiatic Imperial Rivalries, *Modern Asian Studies* 45,

no. 6 (November 2011): 1337–1382. Kairanwi's last name is not always transliterated in this form; spelling varies.

56. Hanssen, *Fin de siècle,* 166, 170, 174–175; Elshakry, *Reading Darwin,* 136.
57. "Qimat al-Waqt," *Thamarat al-Funun,* April 29, 1908, 2–3, here 2.
58. Ibid., 3.
59. For similar thoughts on the role that every individual is assigned in life depending on his capacities, see Musa Saydah, "Al-Waqt," *al-Hilal* 4, no. 3 (October 1, 1895): 93–95, here 94.
60. "Qimat al-Waqt," *Thamarat al-Funun,* April 29, 1908, 3.
61. On *dahr,* see Dalya Cohen Mor, *A Matter of Fate: The Concept of Fate in the Arab World as Reflected in Modern Arabic Literature* (Oxford: Oxford University Press, 2001); W. Montgomery Watt, "Dahr," EI2, Brill Online, July 12, 2014, http://referenceworks.brillonline.com/entries/encyclopaedia-of-islam-2/dahr-SIM_1665?s.num=0&s.f.s2_parent=s.f.book.encyclopaedia-of-islam-2&s.q=dahr+montgomery+watt. *Dahr* is not identical with Dahriyya, which is often translated as "materialism." Jamal al-Din al-Afghani himself wrote a famous tract against materialist philosophy, his *Refutation of the Materialists.* See Ignaz Goldziher and A. M. Goichon, "Dahriyya," EI2, Brill Online, July 12, 2014, http://referenceworks.brillonline.com/entries/encyclopaedia-of-islam-2/dahriyya-SIM_1666?s.num=0&s.f.s2_parent=s.f.book.encyclopaedia-of-islam-2&s.q=goldziher+dahriyya. On the meaning of *dahr* and *zamān* in Islamic philosophy, see Muhammad al-Rahmuni, *Mafhum al-Dahr fi al-'Alaqa Bayn al-Makan wa-l-Zaman fi al-Fada' al-'Arabi al-Qadim* (Beirut: al-Shabaka al-'Arabiyya Li-l-'Abhath wa al-Nashr, 2009).
62. Cohen Mor, *Matter of Fate,* 47; Albert Hourani, *Arabic Thought in the Liberal Age, 1798–1939* (Cambridge: Cambridge University Press, 1983), 183.
63. As'ad Daghir, "Al-Shabab wa-l-Waqt," *al-Muqtataf* 14, no. 6 (March 1, 1890): 384–385, here 384.
64. See As'ad Daghir, "Al-Shabab wa-l-Waqt: Tabi' ma Qablahu," *al-Muqtataf* 14, no. 10 (July 1, 1890): 668–670, here 668.
65. Ibid., 669.
66. Ibid., 670.
67. As'ad Daghir, "Al-Shabab wa-l-Waqt: Tabi' ma Qablahu," *al-Muqtataf* 15, no. 6 (March 1, 1891): 384–386, here 385.
68. Ibid.
69. Ibid.
70. Ibid., 386.
71. Ma'ruf al-Rusafi, "Al-Sa'a," *al-Muqtabas* 3, no. 6 (May 1908): 279.
72. S. Moreh, "Ma'ruf al-Rusafi," EI2, Brill Online, July 12, 2014, http://referenceworks.brillonline.com/entries/encyclopaedia-of-islam-2/maruf-al-rusafi-SIM_4976?s.num=0&s.f.s2_parent=s.f.book.encyclopaedia-of-islam-2&s.q=moreh+rusafi.
73. Al-Rusafi, "Al-Sa'a," 279.
74. Salim Diyab, "Mulaqat al-Zaman," *al-Jinan,* no. 19 (October 1, 1875): 665–669, here 666.

75. Ibid., 667.
76. Ibid.
77. Both "progress" and "backwardness" and "setting forward" and "setting backward" are derived from the same root in Arabic: taqaddum and ta'akhkhur, and taqdīm and ta'khīr. On Arab historiography in the late nineteenth and early twentieth century, see Yoav Di-Capua, *Gatekeepers of the Arab Past: Historians and History Writing in Twentieth-Century Egypt* (Berkeley: University of California Press, 2009), ch. 1; Werner Ende, *Arabische Nation und Islamische Geschichte. Die Umayyaden im Urteil arabischer Autoren des 20. Jahrhunderts* (Wiesbaden: Franz Steiner Verlag, 1977), esp. ch. III.
78. Diyab, "Mulaqat al-Zaman," 668.
79. Elshakry, *Reading Darwin*, 11–12.
80. "Akhbar Wataniyya," *al-Muqtataf* 5, no. 1 (June 1880): 29–30.
81. "Shahhada fi Thuraya Falakiyya Tushakhkhis Nizam Dawrat al-'Ard, Ikhtara'aha al-Khawaja Ilyas Ajiya," *Thamarat al-Funun*, January 26, 1881, 3.
82. Ibid.; on Ajiya's visit to Paris, see "Al-Sa'a al-Falakiya," *Al-Tabib* (June 15, 1884): 2; "Séance du 6 juillet 1883," *Comptes rendus des séances de la Société de Géographie et de la commission centrale, anné 1882* (Paris: Société de Géographie, 1883): 371–373.
83. On Smiles, see Adrian Jarvis, *Samuel Smiles and the Construction of Victorian Values* (Thrupp, Gloucestershire: Sutton, 1997); *Tim Travers, Samuel Smiles, and the Victorian Work Ethic* (New York: Garland, 1987).
84. Samuel Smiles, *Self-Help. With Illustrations of Character and Conduct* (London, 1859), 199–200.
85. Smiles, *Self-Help*, 1–2. At the top of the page, Smiles provided the reference to his own source for this quote, John Stuart Mill, who wrote, "the worth of the state, in the long run, is the worth of the individuals composing it." See John Stuart Mill, *On Liberty*, 2nd ed. (London: John W. Parker and Son, 1859), 207.
86. Samuel Smiles, *The Autobiography of Samuel Smiles* (London: J. Murray, 1905), 230; Timothy Mitchell, *Colonizing Egypt* (Berkeley: University of California Press, 1988), 108.
87. Jurji Zaydan, *Mudhakkirat Jurji Zaydan* (Beirut, 1968). The translation is Thomas Philipp's from *The Autobiography of Jurji Zaidan* (Washington, DC: Three Continents Press, 1990), 44, 69.
88. National Library of Scotland, John Murray Archives (hereafter NLS), Acc12927/285/H8 Samuel Smiles, general file, letter, Smiles, February 29, 1912. Some translations were made slightly later in the 1920s, and in exceptional cases such as the Persian one, in 1933, but the bulk seems to fall in the second half of the nineteenth century.
89. NLS Acc12927/285/H8 Samuel Smiles, general file, letter, Raj Kumar to John Murray, Dacca, March 4, 1903; Samuel Smiles, general file, letter, B. Banerjee of Banerjee & Co. Booksellers and Publishers to John Murray, Calcutta, November 23, 1911; NLS John Murray, MS 41099, Folio 173, letter, Samuel Smiles to John Murray, March 8, 1869; see also MS 42203, letter, Smiles to

Murray, May 11, 1877; see also Acc12927/285/H8 Samuel Smiles, general file, letter, Smiles, February 29, 1912, with a list of existing translations authorized by Smiles or Murray by 1912. In addition to the several authorized translations and partial translations, numerous unauthorized translations existed.

90. Donald M. Reid, "Syrian Christians, the Rags-to-Riches Story, and Free Enterprise," *International Journal of Middle East Studies* 1, no. 4 (1970): 358–367, esp. 362.

91. On the notion of the self-made man in Japan, see Earl H. Kinmonth, *The Self-Made Man in Japanese Thought: From Samurai to Salary Man* (Berkeley: University of California Press, 1981); see also Earl H. Kinmonth, "Nakamura Keiu and Samuel Smiles: A Victorian Confucian and a Confucian Victorian," *American Historical Review* 85, no. 3 (1980): 535–556. Most recently, Sheldon Garon has compared, among other things, the education to thrift that Smiles promoted as a transnational history between Britain, Japan, and the United States: Sheldon Garon, *Beyond Our Means: Why America Spends While the World Saves* (Princeton, NJ: Princeton University Press, 2012). Fukuzawa Yukichi who translated *Self-Help* into Japanese also penned an essay on the defense of the Gregorian calendar when the latter was introduced in Japan in 1873. In it, Yukichi furthermore explained how a Western watch counting twelve hours and having two clock hands should be read "clockwise" from right to left; Florian Coulmas, *Japanische Zeiten: Eine Ethnographie der Vergänglichkeit* (Reinbek/Hamburg: Kinder, 2000), 122. See also Nishimoto Ikuko, "The 'Civilization' of Time: Japan and the Adoption of the Western Time System," *Time and Society* 6 (1997): 237–259; and Stefan Tanaka, *New Times in Modern Japan* (Princeton, NJ: Princeton University Press, 2004).

92. In a wonderful book, On Barak analyzes such critiques and occasionally subversions of European time based on a close reading of countless newspapers, magazines, and other cultural productions from Egypt; see Barak, *On Time: Technology and Temporality in Modern Egypt* (Berkeley: University of California Press, 2013).

93. Ibrahim Ramzi, *Kitab Asrar al-Najah* (Cairo, 1911), for instance 32, 35, 63–84, and especially the sections on trade.

## 6. Islamic Calendar Times

1. Other notorious controversies over matters of science concerned Darwin's theories and Copernicanism; for example, see Elshakry, *Reading Darwin*; Stolz, *Lighthouse*, esp. 233. On the adoption of scientific thought in Egypt, see Pascal Crozet, *Les sciences modernes en Egypte: Transfert et appropriation 1805–1902* (Paris: Geuthner, 2008). In part these controversies were mirrored by struggles over the curriculum taught at the Islamic world's foremost institution of higher learning, Cairo's al-Azhar University. See Jakob Skovgaard-Petersen, *Defining Islam for the Egyptian State: Muftis and Fatwas of the Dār al-Iftā'* (Leiden: Brill, 1997), 49.

2. On the Islamic calendar, see Stowasser, *Day Begins at Sunset*, ch. 2.

3. Jamal al-Din al-Qasimi, "Bushra 'Ilmiyya," *al-Muqtabas,* October 18, 1910/12 Shawwal 1328, 1; Skovgaard-Petersen, *Defining Islam,* 87.
4. Jamal al-Din al-Qasimi, *Irshad al-Khalq Ila al-'Amal bi-Khabar al-Barq* (Damascus: Matba'a al-Muqtabas, 1911), 93.
5. Muhammad 'Arif al-Munayyir al-Husayni, "Munazara Bayn 'Alimayn (pt. II)," *al-Haqa'iq* 1 (1910): 264–267, here 265. On Bakhit, see Junaid Quadri, "Transformations of Tradition: Modernity in the Thought of Muhammad Bakhit al-Muti'i" (PhD diss., McGill University, 2013).
6. Al-Qasimi, *Irshad al-Khalq,* 93.
7. Ibid., 71.
8. Al-Munayyir, "Munazara (pt. II)," 265.
9. Ibid.
10. Bakhit al-Muti'i, *Irshad 'Ahl al-Milla,* 16.
11. Wael Hallaq, *An Introduction to Islamic Law* (Cambridge: Cambridge University Press, 2009), 8.
12. Ibid., 10–11; Skovgaard-Petersen, *Defining Islam,* 7.
13. This was a period when moreover, at the same time, the shari'a courts and state juridical personnel were undergoing reform, at times enlisting staunch opposition and conflict among the 'ulamā'. See Skovgaard-Petersen, *Defining Islam,* 60–63.
14. Wael Hallaq, *A History of Islamic Legal Theory: An Introduction to Sunnī 'Uṣūl al-Fiqh* (Cambridge: Cambridge University Press, 1997), 14.
15. Ibid., 27.
16. What remained open to debate was whether it was mandatory to act and thus rule on the basis of a solitary message, or whether it was merely permissible to do so. Most authors discussed in this chapter differed on the exact implications of action and permissibility. On the epistemology of certainty in Islamic legal theory, see Aron Zysow, *The Economy of Certainty: An Introduction to the Typology of Islamic Legal Theory* (Atlanta: Lockwood Press, 2013). My thanks to Ahmed El Shamsy for the reference.
17. Skovgaard-Petersen, *Defining Islam,* 84.
18. David Dean Commins, *Islamic Reform: Politics and Social Change in Late Ottoman Syria* (New York: Oxford University Press, 1990), 4. See also Itzchak Weismann, *Taste of Modernity: Sufism, Salafiyya, and Arabism in Late Ottoman Damascus* (Leiden: Brill, 2001).
19. Commins, *Islamic Reform,* 3.
20. Al-Qasimi, *Irshad al-Khalq,* 12.
21. On the formation of the four main schools *(ḥanbalī, ḥanafī, mālikī, shāfi'ī)* in Sunni Islam, see Ahmed El Shamsy, *The Canonization of Islamic Law: A Social and Intellectual History* (Cambridge: Cambridge University Press, 2013).
22. Commins, *Islamic Reform,* 119.
23. "Mas'alat al-Tilighraf wa Khulasat al-Qawl Fiha," *al-Haqa'iq* 1 (1910): 171–182, here 174.
24. Ibid., 175.
25. Ibid.

26. Al-Munayyir, "Munazara Bayn 'Alimayn (pt. II)," 265. Another known rejection of the telegraph did not elaborate on the legal implications of using the telegraph for determining the lunar month but rather declared telegraphy to lack legal validity in general. See Rudolph Peters, "Religious Attitudes towards Modernization in the Ottoman Empire: A Nineteenth-Century Pious Text on Steamships, Factories, and the Telegraph," *Die Welt des Islam* 26, no. 1/4 (1986): 76–105, here 93.
27. Al-Munayyir, "Munazara Bayn 'Alimayn (pt. II)," 267. *Al-Haqa'iq* published two more attacks on al-Qasimi: Muhammad 'Arif al-Munayyir al-Husayni, "Munazara Bayn 'Alimayn," *al-Haqa'iq* 1 (1910): 211–216, and Mukhtar al-Mu'ayyad [al-'Azm], "'Awd 'ala Bud'," *al-Haqa'iq* 1 (1910): 310–313. See also Commins, *Islamic Reform,* 173.
28. The authors of these fatwas were Muhammad Sa'id, then mufti of Algiers; Muhammad Bakhit "al-Muti'i"; Khalil Hammad "al-Luddi" of Palestine; Muhammad Rashid Rida; 'Abd al-Baqi "al-Afghani"; Muhammad al-Sadiq "al-Shatti" (d. 1889) of Damascus; 'Abd al-Razzaq al-Bitar (1837–1917) of Damascus; Salim al-Bishri of Cairo; Muhammad 'Abbasi "al-Mahdi" (1827–1897) of Cairo (on al-'Abbasi, see Skovgaard-Petersen, *Defining Islam,* 106–111); Muhammad Kamil "al-Trabulsi" of Tripoli in Libya; Muhammad Ibn Ahmad 'Ullaysh (1802–1882; sometimes transliterated as 'Illish) of Cairo. The poems are by a Lebanese historian by the name of Issa Iskandir Ma'luf, the other by a Lebanese shi'ite cleric from the southern Lebanese region of Jabal 'Amil, 'Abd al-Hussayn Sadiq, "a preeminent religious figure" in the southern city of Nabatiyya. Max Weiss, *In the Shadow of Sectarianism: Law, Shi'ism, and the Making of Modern Lebanon* (Cambridge, MA: Harvard University Press, 2010), 74. Another salafi reformer who opined on timekeeping inspired by telegraphy was Mahmud Shukri al-Alusi. See Mahmud Shukri al-Alusi, *Ma Dalla 'alayhi al-Qur'an mimma Ya'dud al-Hay'a al-Jadida al-Qawimat al-Burhan* (Damascus: al-Maktab al-Islami, 1960). On al-Alusi, see Itzchak Weismann, "Genealogies of Fundamentalism: Salafi Discourse in Nineteenth-Century Baghdad," *British Journal of Middle Eastern Studies* 36, no. 2 (2009): 267–280.
29. Al-Qasimi, *Irshad al-Khalq,* 72.
30. David Commins translated the first ten pages of al-Qasimi's book for inclusion in Charles Kurzman's sourcebook for Middle East history. See "Jamal al-Din al-Qasimi: Guiding Mankind to Act on the Basis of Telegraphic Messages," in *Modernist Islam, 1840–1940: A Sourcebook,* ed. Charles Kurzman (Oxford: Oxford University Press, 2002): 181–187, here 182–183.
31. Sura 16, verse 8; quoted after Commins, "Jamal al-Din al-Qasimi," 183.
32. Eugene Rogan, "Instant Communication: The Impact of the Telegraph Network in Ottoman Syria," in *The Syrian Land: Processes of Integration and Fragmentation: Bilād al-Shām from the 18th to the 20th Century,* ed. Thomas Philipp and Birgit Schäbler (Stuttgart: Steiner, 1998), 113–118, here 115–116; Yakup Bektas, "The Sultan's Messenger: Cultural Constructions of Ottoman Telegraphy, 1847–1880," *Technology and Culture* 41, no. 4 (2000): 669–696; Barak, *On Time,* 44–45.
33. Al-Qasimi, *Irshad al-Khalq,* 22.
34. Ibid., 67.

35. Ibid., 55. Another scholar had previously argued in a similar vein, contending that such measures functioned as a guarantee against falsification. See "Al-'Amal Bi Khabar al-Tilighraf wa-l-Tilifun," *al-Manar* 7, no. 18 (16 Ramadan 1322/ November 24, 1904): 697.
36. Al-Qasimi, *Irshad al-Khalq,* 68.
37. Ibid., 57.
38. Ibid., 23, 24, and with a different explanation, 31.
39. Ibid., 24.
40. Ibid., 99.
41. Ibid.
42. Ibid., 52.
43. Ibid., 19.
44. Ibid., 59, 64; Bakhit al-Muti'i, *Irshad Ahl al-Milla,* 151.
45. Al-Qasimi, *Irshad al Khalq,* 67; Bakhit al-Muti'i, *Irshad Ahl al-Milla,* 152.
46. Bakhit al-Muti'i, *Irshad Ahl al-Milla,* 17. When it was printed, a famous fifteenth-century tract on moon-sighting and mathematical astronomy by Taqi al-Din al-Subki was attached to Bakhit's book. Al-Subki's views on astronomy still had currency among Muslim astronomers in the early twentieth century.
47. Al-Qasimi, *Irshad al-Khalq,* 93.
48. Bakhit al-Muti'i, *Irshad Ahl al-Milla,* 17. Normally, that shaykh should have been Maulana 'Abd al-Hayy "al-Luknawi" (1848–1886), but given that he passed away quite some time before Bakhit was writing his account of the events, it is unsure whether he was really the one the Saudi friend encountered on his journey.
49. Bakhit al-Muti'i, *Irshad Ahl al-Milla,* 17.
50. Al-Qasimi, *Irshad al-Khalq,* 95; Bakhit al-Muti'i, *Irshad Ahl al-Milla,* 162–163; Skovgaard-Petersen, *Defining Islam,* 88.
51. Commins, *Islamic Reform,* 47.
52. Muhammad 'Abd al-Baqi al-Afghani, *Kitab al-Fawa'id al-Nafi'at Fi Ahkam al-Silk wa-l-Sa'at* (Damascus? s.n., 1897/1898), 5.
53. David A. King, "Science in the Service of Religion: The Case of Islam," in *Astronomy in the Service of Islam,* ed. David A. King (Aldershot, UK: Ashgate, 1993), 245–262, here 249.
54. Ibid., 250; Stowasser, *Day Begins at Sunset,* 148.
55. 'Abd al-Baqi, *Kitab al-Fawa'id,* 6.
56. Ibid., 7.
57. Ibid., 8.
58. Ibid., 9.
59. Ibid., 10.
60. Ibid., 11.
61. Stephen Blake, *Time in Early Modern Islam: Calendar, Ceremony, and Chronology in the Safavid, Mughal, and Ottoman Empires* (Cambridge: Cambridge University Press, 2013), viii.
62. A. J. Wensinck, "Mikat," EI2, BrillOnline, http://referenceworks.brillonline.com/entries/encyclopaedia-of-islam-2/mikat-COM_0735?s.num=0&s.f.s2_parent=s.f.book.encyclopaedia-of-islam-2&s.q=mikat+wensinck.

63. King, "Science in the Service of Religion," 252; Stowasser, *Day Begins at Sunset,* 155.
64. King, "Science in the Service of Religion," 247.
65. Ibid., 249.
66. Ahmad Musa al-Zarqawi, *Kitab 'Ilm al-Miqat: Muqarrar Talabat al-Azhar wa-l-Ma'ahid al-'Ilmiyya 'ala Muqtada al-Namuzaj alladhi Qarrarahu Majlis al-Azhar al-'Ali* (Cairo: Matba'at al-Hilal, 1912).
67. Commins, *Islamic Reform,* 61.
68. "Fasl fima Yuthbat bihi al-Sawm wa-l-Fitr," *al-Manar* 6, no. 17 (1 Ramadan 1321/November 20, 1903): 814–816, here 814. The pagination in this reference is from the first printing of the journal.
69. Ibid., 814–815.
70. Ibid., 815.
71. "Ithbat Ramadanina hadha fi Misr," *al-Manar* 7, no. 18 (16 Ramadan 1322/November 24, 1904): 697–698, here 698.
72. Ibid.
73. "Ra'y Mashayikh al-'Asr fi Dhalik," *al-Manar* 7, no. 18 (16 Ramadan 1322/November 24, 1904): 699–701, here 699. See also another subsection titled "Tariqat Ithbat Ramadan fi Amsar al-Muslimin," *al-Manar* 7, no. 18 (16 Ramadan 1322/November 24, 1904): 695–696; and "Al-'Amal bi Hisab al-Hasibin fi al-Ibada," *al-Manar* 7, no. 18 (16 Ramadan 1322/November 24, 1904): 698–699.
74. "Ra'y Mashayikh al-'Asr fi Dhalik," *al-Manar* 7, no. 18 (16 Ramadan 1322/November 24, 1904): 699–701, here 699.
75. Ibid. On the ambivalences of the countryside episode, see Stolz, *Lighthouse,* 308–309.
76. "Fatawa al-Manar: Su'al 44 'an Hilal al-Sawm wa-l-Fitr min Suakin (al-Sudan)," *al-Manar* 10, no. 7 (30 Rajab 1325/September 8, 1907): 530–531, here 530. Rida's answer followed as "Fasl fima Yuthbat bihi al-Sawm wa-l-Fitr," *al-Manar* 10, no. 7 (30 Rajab 1325/September 8, 1907): 531–534.
77. "Fatawa al-Manar: Al-'Amal bi-Khabar al-Tilifun wa-l-Tilighraf fi al-Sawm wa-l-Fitr," *al-Manar* 13, no. 3 (30 Rabi' al-Awwal 1329/April 10, 1910): 187–190, here 188.
78. Ibid. Rida printed a fatwa by Salim al-Bishri with his answer, the same author and Azhar mufti whose verdict al-Qasimi had cited.
79. "Fatawa al-Manar: As'ila min Madinat Bankuk (Siyam)," *al-Manar* 23, no. 8 (29 Safar 1341/October 20, 1922): 584–585, here 584; and Rida's answer: "Jawab al-Manar: Ithbat Hilal Ramadan wa-l-'Idayn," *al-Manar* 23, no. 8 (29 Safar 1341/October 20, 1922): 585–588.
80. "Mas'alat Hilal Ramadan," *al-Manar* 31, no. 4 (30 Jumada al-Uwla 1349/October 22, 1930): 278.
81. "I'tibar Ru'yat al-Hilal fi al-Shuhur al-'Arabiyya: Min Rada' al-Din Effendi, Qadi al-Quda' fi Ufa (al-Rusiya)," *al-Manar* 6, no. 18 (16 Ramadan 1321/December 5, 1903): 705–707, here 706.
82. Ibid.
83. Ibid., 707.

84. See Ordinance #8 dating from 1923 quoted in Mahmud Naji, *Natijat al-Dawla al-Misriyya li-Sanat 1354 Hijriyya* (Cairo: al-Matbaʿa al-ʾAmiriyya, 1935), 13.
85. Stolz, *Lighthouse,* 293.
86. Hallaq, *History,* 216.
87. In at least one instance Rida seems to have approved of calculation, under very specific circumstances: for an analysis, see Stolz, *Lighthouse,* 330–331.
88. ʿAbd al-Baqi al-Afghani, *Kitab al-Fawaʾid,* 12.
89. Al-Qasimi, *Irshad al-Khalq,* 87. The fatwa under concern was by Salim al-Bishri.
90. Ibid., 49.
91. "Fasl fima Yuthbat bihi al-Sawm wa-l-Fitr," 815.
92. Ibid., 816.
93. Moosa, "Shaykh Ahmad Shākir," 59.
94. Ibid., 65, 77. On al-Maraghi, see Skovgaard-Petersen, *Defining Islam,* 160–161.
95. Moosa, "Shaykh Ahmad Shākir," 69.
96. Ibid., 70.
97. Ibid., 70, 81.
98. Ibid., 81.
99. Ibid., 86.
100. "Fasl fima Yuthbat bihi al-Sawm wa-l-Fitr," 816. "Great Imam," especially *the* Greatest Imam [*al-Imām al-Aʿẓam*], can also refer to Abu Hanifa, the founder of the ḥanafī school of Islamic law. But given the contextual information provided by Rida, caliph is a more likely translation.
101. On Islamic internationalism, see Martin Kramer, *Islam Assembled: The Advent of the Muslim Congresses* (New York: Columbia University Press, 1986); Reinhard Schulze, *Islamischer Internationalismus im 20. Jahrhundert: Untersuchungen zur Geschichte der Islamischen Weltliga (Rābiṭat al-ʿĀlam al-Islāmī) Mekka* (Leiden: Brill, 1990).
102. Among those calling for calculation around the middle decades of the twentieth century were Abu al-Nasr Mubashshir al-Tirazi of Taraz (presently Kazakhstan) and temporarily Bukhara (presently Uzbekistan). See Moosa, "Shaykh Ahmad Shakir," 65, and Bakr Ibn ʿAbd Allah Abu Zayd, "Tawhid Bidayat al-Shuhur al-Qamariya," *Majallat Majmaʿ al-Fiqh al-Islami: al-Dawra al-Thalitha li-Muʾtamar Majmaʿ al-Fiqh al-Islami* 3, no. 2 (1408/1987): 820–841, here 827.
103. Moosa, "Shaykh Ahmad Shakir," 66–67. Saudi Arabia today uses the astronomically calculated *ʾumm al-qurā* calendar for administrative purposes but retains the practice of sending out teams of moon-sighters to actually spot the crescent.
104. Anderson, *Imagined Communities;* Nile Green, "Spacetime and the Muslim Journey West: Industrial Communications in the Makings of the 'Muslim World,'" *American Historical Review* 118, no. 2 (April 2013): 401–429.
105. On imagining a global Muslim ʾumma, see Cemil Aydin, "Globalizing the Intellectual History of the Idea of the 'Muslim World,'" in *Global Intellectual*

*History,* ed. Samuel Moyn and Andrew Sartori (New York: Columbia University Press, 2013), 159–186.

106. On the adoption and reconfiguration of Western science in China "on their own terms," see Benjamin A. Elman, *On Their Own Terms: Science in China, 1550–1900* (Cambridge, MA: Harvard University Press, 2005).

## 7. One Calendar for All

1. As a popularizer of science, Flammarion embodied the fluid boundaries between science and what today would be called pseudoscience. Besides astronomy, he wrote about extraterrestrial life. On pseudoscience, see Michael Gordin, *The Pseudoscience Wars: Immanuel Velikovsky and the Birth of the Modern Fringe* (Chicago: University of Chicago Press, 2012).
2. Camille Flammarion, "Les imperfections du calendrier, projet de réforme par M. Camille Flammarion," *Bulletin de la société astronomique de France et revue mensuelle d'astronomie, de météorologie et de physique du globe* 14 (1901): 311–327, here 311; "The Reform of the Calendar," *Nature,* December 25, 1919, 415–416.
3. Flammarion, "Les imperfections du calendrier," 313.
4. Ibid., 314; see also Alexander Philip, *The Reform of the Calendar* (London: Kegan Paul, Trench, Truebner, & Co., 1914), 58–59; and Alexander Philip, *The Improvement of the Gregorian Calendar* (London: George Routledge & Sons, 1918).
5. "Société astronomique de France, séance du 5 juin 1901," *Bulletin de la société astronomique de France et revue mensuelle d'astronomie, de météorologie et de physique du globe* 14 (1901): 300–309, here 300, 305.
6. Wilhelm Förster, *Kalenderwesen und Kalenderreform* (Braunschweig: Friedrich Vieweg & Sohn, 1914).
7. Josef Jatsch, *Kalenderreform und Völkerbund, an Stelle der Feierlichen Inauguration des Rektors der Deutschen Universität in Prag für das Studienjahr 1923/1924* (Prague: Self-published by Deutsche Universität Prag, 1928), 21.
8. RGO 9/631, The Calendar and Its Reform, Text of Address by Sir Harold Spencer Jones of London (Astronomer Royal) at the American Museum-Hayden Planetarium, December 4, 1953.
9. "A New Calendar Clock," *Nature,* May 8, 1879, 35–36.
10. Mitchell, *Rule of Experts.*
11. BArch R901/37729, "Kalenderreform in China," *Tägliche Rundschau,* September 15, 1911; "Time in Turkey," *Times of India,* February 20, 1909, 14.
12. Heinrich von Mädler, "Die Kalender-Reform mit spezieller Beziehung auf Russland (1864)," in von Mädler, *Reden und Abhandlungen über Gegenstände der Himmelskunde* (Berlin: Robert Oppenheim, 1870), 350–355.
13. On Tondini's early years, see Luca Carboni, "Cesare Tondini, Gli anni della giovinezza: 1839–1871 (formazione, missione e primi scritti)," *Studi Barnabiti* 22 (2005): 95–195.

14. BArch R901/37725, letter, Rome, June 7, 1894; BArch R 901/37725, Wilhelm Förster, Schrift über Anregungen betreffend Reform des Osterfestes und Unifikation des Kalenders, Berlin, April 22, 1894; Tcheng-Ki-Tong, "La Turquie: Le Calendrier Universel et le Méridien Initial de Jerusalem," *La nouvelle revue* 10, no. 55 (November/December 1888): 440–443, here 443; "The British Association," *Times of London,* September 13, 1888; Cesario Tondini de Quarenghi, "The Gregorian Calendar and the Universal Hour," *Times of London,* October 2, 1888, 13; Russologus, "Le Calendrier Russe," *La nouvelle revue* 10, no. 54 (September/October 1888): 834–842, here 841.
15. BArch R901/37729, newspaper clipping "Russische Reformbestrebungen," *National Zeitung,* February 3, 1910.
16. "Difficulties of the Calendar," *Nature,* March 22, 1900, 493–494; BArch R 901/37729, draft letter, Berlin, January 23, 1911, 8–12; on calendars generally, see Anthony Aveni, *Empires of Time: Calendars, Clocks, and Cultures* (New York: Basic Books, 2000); BArch R 901/37727, newspaper clipping "La Question des calendriers," *Levant Herald,* August 6, 1904.
17. BArch R901/37729, "Ein Welt-Kalender?," *Kölnische Zeitung,* October 10, 1910.
18. Ibid.
19. Ibid.
20. See RGO 9/631, International World Calendar Association to Harold Spencer Jones, April 3, 1956.
21. E. G. Büsching, *Die Kalenderreform: Einführung einer feststehenden, von Jahr zu Jahr gleichmässigen und möglichst regelmässigen Jahreseinteilung* (Halle: Verlag Rudolf Heller, 1911), 28.
22. BArch R901/37725, letter, Ignatz Heising to Ministry of Foreign Affairs, Pittsburg, received July 3, 1893.
23. Wilhelm Nehring, *Die internationale Handelskammer: Ihre Geschichte, Organisation und Tätigkeit* (Jena: G. Neuenhahn GmbH, 1929), 11.
24. "An International Movement to Change the Calendar," *New York Times,* October 6, 1912, 8.
25. Madeleine Herren, *Hintertüren zur Macht: Internationalismus und modernisierungsorientierte Aussenpolitik in Belgien, der Schweiz und den USA 1865–1914* (München: Oldenburg, 2000), 161–164.
26. BArch R901/37729, newspaper clipping "Kalender-Reform," *Frankfurter Zeitung,* July 17, 1910; BNA HO 45/10548/162178, draft, Calendar Reform Act 1908, An Act to Reform the Calendar Fixing Easter and Other and More Bank Holidays and for Other Purposes in Relation Thereto, 1.
27. BNA HO 45/10548/162178, draft, Calendar Reform Act 1908, An Act to Reform the Calendar Fixing Easter and Other and More Bank Holidays and for Other Purposes in Relation Thereto, 2; see also Philip, *Reform of the Calendar,* 58; "Britain Asks Views on Calendar Reform," *New York Times,* March 31, 1911, 2.
28. BArch R901/37726, memorandum Aufzeichnung über die Frage der Einführung eines Normalkalenders und der Festlegung des Osterfestes, Berlin, Jan-

uary 23, 1911; "Proposals for the Reform of the Calendar," *Nature,* August 26, 1911, 281–283.

29. League of Nations Archives, Communications and Transit Section (hereafter quoted as LNA CTS) Calendar Reform, 1919–1927, 14/22679/12478, Reform of the Calendar, Provisional Committee, Document 8, Note on the Experience of Reforming the Calendar, March 9, 1922, 1.
30. On the history of the League generally as well as the Communications and Transit Section, see Francis P. Walters, *A History of the League of Nations* (Oxford: Oxford University Press, 1960).
31. LNA CTS, Calendar Reform, 1919–1927, 14/22679/12478, Reform of the Calendar, Provisional Committee, Document 8, Note on the Experience of Reforming the Calendar, March 9, 1922, 2.
32. LNA CTS, 1919–1927, 14/52722/12478, Resolution 1926, Société des Nations, commission consultative et technique des communications et du transit, réforme du calendrier, projet de résolution, July 17, 1926, 1.
33. Sarah Steiner, *The Lights that Failed: European International History, 1919–1933* (Oxford: Oxford University Press, 2005).
34. Pamphlet, The World Calendar Association, Calendar before the United Nations, Basic Facts, n.d., 4.
35. Charles S. Maier, *Recasting Bourgeois Europe: Stabilization in France, Germany, and Italy in the Decade after World War I* (Princeton, NJ: Princeton University Press, 1975).
36. BArch R1501/125214, Report of the National Committee on Calendar Simplification for the United States, Submitted to the Secretary of State (Rochester, NY: Office of the Chairman, 1929), 11.
37. Hirosi Saito, "Japan's Attitude," *Journal of Calendar Reform* 4, no. 4 (1934): 106. See also Benjamin C. Gruenberg, "Time in Our Changing Time," *Journal of Calendar Reform* 8, no. 2 (1938): 87–90, here 87.
38. League of Nations Publication No. C.977.M.542, Organization for Communication and Transit, Records and Texts Relating to the Fourth General Conference on Communications and Transit, Geneva, October 12–24, 1931, vol. 1, annex 9, 71; Rudolf Blochmann, "Die Kalenderreform, ein Hilfsmittel zur Förderung des Völkerfriedens," *Friedens-Warte* 30, no. 3 (1930): n.p. See also the prewar volumes of the *Journal of Calendar Reform* from 1938 and 1939.
39. Alfred D. Chandler, *The Visible Hand: The Managerial Revolution in American Business* (Cambridge, MA: Belknap Press of Harvard University Press, 1977); on accountancy see also Jacob Soll, *The Reckoning: Financial Accountability and the Rise and Fall of Nations* (New York: Basic Books, 2014).
40. Chandler, *Visible Hand,* 297.
41. "Says Business Men Back 13-Month Year," *New York Times,* October 26, 1928, 28; see also "Business Men Urge Calendar Reform: George Eastman Tells House Committee that 1929 Is Year for Parley on 13-Month Plan," *New York Times,* December 21, 1928, 9.
42. LNA CTS, 1919–1927, 14/36384x/12478, Reform of the Calendar, Advisory and Technical Committee for Communications and Transit, draft minutes of

the first meeting of the committee, May 19, 1924, 4; H. Parker Willis, "Economic Aspects," *Journal of Calendar Reform* 1, no. 2 (1931): 34. Quoted after H. Parker Willis, "Another Financial Fallacy," *Journal of Calendar Reform* 3, no. 1 (1933): 19–26, here 20.

43. BArch R1501/125215, Annex, International Association of Railways, Committee for Accounting and Currencies, Subcommittee for International Railway Statistics, Calendar Reform, 5–6.
44. Ibid., 7.
45. Ibid., 12. Earlier, US and British railways were said to use thirteen-month accounting schemes; see Jatsch, *Kalenderreform und Völkerbund,* 21.
46. BArch R1501/125214, American Committee on Calendar Reform: Present Status of the Special Committee, 1.
47. BArch R1501/125214, Report of the National Committee on Calendar Simplification for the United States, Submitted to the Secretary of State (Rochester, NY: Office of the Chairman, 1929), 21. The committee received a total of 1,433 completed questionnaires, out of which 488 had indicated a preference for either one of the plans, with 480 of those preferences favoring the thirteen-month plan.
48. Benedict Anderson, *The Specter of Comparison: Nationalism, Southeast Asia, and the World* (London: Verso, 1998), 36. On indicators, see Zachary Karabell, *The Leading Indicators: A Short History of the Numbers that Rule Our World* (New York: Simon & Schuster, 2013).
49. F. L. S. Lyons, *Internationalism in Europe, 1815–1914* (Leiden: A. W. Sythoff), 125. On the early history of this "international congress of statisticians," see Jacques Dûpaquier and Michel Dûpaquier, eds., *Histoire de la démographie: la statistique de la population des origines à 1914* (Paris: Libr. Académique Perrin, 1985), esp. 309; Alain Désrosières, "Histoire de la statistique: styles d'ecriture et usages sociaux," in *The Age of Numbers: Statistical Systems and National Traditions,* ed. Jean-Pierre Beaud and Jean-Guy Prevost (Quebec: Presses de l'Université du Québec, 2000), 37–57; Marc-André Gagnon, "Les réseaux de l'internationalisme statistique (1885–1914)," in Beaud and Prevost, *Age of Numbers,* 189–219.
50. On accounting, statistics, and World War I, see Anne Loft, *Understanding Accounting in Its Social and Historical Context: The Case of Cost Accounting in Britain, 1914–1925* (New York: Garland Publishing, 1988), 147. On forecasting, see Walter Friedman, *Fortune Tellers: The Story of America's First Economic Forecasters* (Princeton, NJ: Princeton University Press, 2014).
51. Adam J. Tooze, *Statistics and the German State, 1900–1945: The Making of Modern Economic Knowledge* (Cambridge: Cambridge University Press, 2001); Daniel Speich Chassé, *Die Erfindung des Bruttosozialprodukts: Globale Ungleichheit in der Wissensgeschichte der Ökonomie* (Göttingen: Vandenhoeck & Ruprecht, 2013); see also BArch R1501/125220, State of Calendar Reform (Result of the Inquiry by the Foreign Office), November 25, 1929, 1; Patricia Clavin, *Securing the World Economy: The Reinvention of the League of Nations, 1920–1946* (Oxford: Oxford University Press, 2013); BArch R1501/125220, Ministry of the Interior to Institute for Business Cycle Research, June 30, 1930.

52. Generally on the role of domesticity and women accompanying male US Foreign Service diplomats, see Molly M. Wood, "Diplomatic Wives: The Politics of Domesticity and the 'Social Game' in the U.S. Foreign Service, 1905–1941," *Journal of Women's History* 12, no. 2 (2005): 142–165 (my thanks to Kristin Hoganson for the reference); "Women in Fascist Ruled States Have Lost Even Normal Rights," *Montreal Gazette,* June 24, 1942, 1; Emma Gelders Sterne, *Blueprints for the World of Tomorrow: A Summary of Peace Plans Prepared for the International Federation of Business and Professional Women, By Emma Gelders Sterne in Collaboration with Marie Ginsberg and Essy Key Rasmussen* (New York: International Federation of Business and Professional Women, 1943).
53. Library of Congress Manuscript Division, World Calendar Association Records (hereafter LOC WCA), series 1: general files, 1931–1955, box 1, newspaper clipping, "She Leads a Crusade: Elisabeth Achelis Backs Calendar That Would Make Every Year Alike," *St. Petersburg Times,* February 21, 1939.
54. LNA CTS, 1928–1932, 9a/25158/1487, Calendar Reform, World Calendar Association, Report of the World Calendar Association, Inc. to the Committee on Communication and Transit of the League of Nations, received June 16, 1931.
55. LOC WCA, series 1: general files, 1931–1955, box 3, Fourth European Trip, May–October 1933, Achelis to Editors of *Journal of Calendar Reform,* August 4, 1933.
56. LNA CTS, 1919–1927, 14/36384x/12478, Draft minutes of the second meeting of the committee of enquiry, May 20, 1924, 1. See also George Eastman, *Do We Need Calendar Reform?* (Rochester, NY: Eastman Kodak Co., 1927), 25; see also *Modification of the Calendar, Hearing Before the Committee on the Judiciary, House of Representatives,* Sixty-Seventh Congress, Second Session, On H. R. 3178, Serial 27, February 9, 1922 (Washington, DC: Government Printing Office, 1922), 15; Moses B. Cotsworth, *The Rational Almanac: Tracing the Evolution of Modern Almanacs from Ancient Ideas of Time, and Suggesting Improvements* (Acomb, UK: Self-published, 1902).
57. LNA CTS, 1928–1932, 9a/1487/1487, Calendar Reform, Correspondence with M. Cotsworth, letter, Haas to Cotsworth, July 26, 1928; and the materials in dossier 9a/14067/1487 on Cotsworth's tour in Latin America; dossier 9a/20987/1487 on a tour in the Far East and Canada; 9a/24495/1487 on Cotsworth's tour in the Near East.
58. LNA CTS, 1928–1932, 9a/24495/1487, Calendar Reform, Cotsworth's tour in the Near East, letter Cotsworth to Haas, January 14, 1931.
59. LNA CTS, 1933–1946, 9a/3743/1174, Calendar Reform, the World Calendar Association, memorandum, Essy Key-Lehmann, The League of Nations and Calendar Reform: Past, Present, and Future Developments, 3.
60. LNA CTS, 1919–1927, 14/22679/12478, Reform of the Calendar, Provisional Committee, League of Nations, Advisory Technical Committee on Communications and Transit, Note on the Reform of Calendar, Geneva, August 14, 1922, 12.
61. *Modification of the Calendar, Hearing before the Committee on the Judiciary,* 41, 5, 16.

62. Albert W. Whitney, *The Place of Standardization in Modern Life, With Introduction by the Honorable Herbert Hoover* (Washington, DC: Inter-American High Commission, 1924). Calendar reform had been brought up previously in a Pan-American context. At the Pan-American Scientific Congress of 1909, Carlos A. Hesse of Peru had presented a thirteen-month calendar: Carlos A. Hesse, *Proyecto de reforma del calendario presentado al 4 congreso cientifico* (Iquique, 1909).
63. *Simplification of the Calendar, Hearings before the Committee on Foreign Affairs,* House of Representatives, Seventieth Congress, Second Session on H. J. Res. 334, A Joint Resolution Requesting the President to Propose the Calling of an International Conference for the Simplification of the Calendar, or to Accept, on Behalf of the United States, an Invitation to Participate in Such a Conference, December 20 and 21, 1928, January 7, 8, 9, 10, 11, 14, 18, and 21, 1929 (Washington, DC: Government Printing Office, 1934), 1.
64. LOC WCA, box 12, folder Central Statistical Board, Lecture transcript, The Use of Four-Week Reporting Periods for Industry with Relation to the Recovery Program, remarks by M. B. Folsom, Chairman of the Committee on Calendar Reform of the American Statistical Association, September 21, 1933; see also The Four-Week Statistical Period, Principal Points in the Discussion at the Meeting of the Committee on Calendar Reform, n.d.
65. Ibid.
66. LNA CTS, 1928–1932, 9a/29580/10782, Reform of the Calendar, India, letter, India Office to Secretary General, London, June 10, 1937; see LNA CTS, 1919–1927, 14/34111x/12478, Reform of the Calendar, India, letter, director of Communications and Transit Section to Overseas Department, February 29, 1924; letter, India Office to Secretary General League of Nations, London, February 20, 1924; 14/36441/12478, Responses of Governments, Extrait du memorandum préparé par l'honorable Diwan Bahadur, January 26, 1924.
67. See Charles D. Morris, "India Launches Reform Movement," *Journal of Calendar Reform* 23 (1953): 59–62, here 60.
68. Ibid., 62; "Nehru Speaks Up for India. Proceedings of Calendar Reform Committee Meeting in Delhi," *Journal of Calendar Reform* 23 (September 1953): 115–121, here 115; Government of India, *Report of the Calendar Committee* (New Delhi: Council of Scientific and Industrial Research, 1955), 1. See also M. N. Saha, "The Reform of the Indian Calendar," *Science and Culture* 18, no. 2 (1952): 57–68, and Saha, "The Reform of the Indian Calendar," *Science and Culture* 18, no. 8 (1953): 355–361.
69. BArch R1501/125221, Adolf Berger to Hitler, Liegnitz, June 20, 1933.
70. BArch R1501/23588, Head of the Office of Imperial Statistics to Ministry of the Interior, April 10, 1935.
71. BArch R1501/23588, Nazi Chamber of Literature to Office of Statistics, Berlin, March 25, 1935; BArch R1501/125218, Ministry of the Interior to all ministries, state governments, Berlin, November 27, 1933; BArch R1501/23588, German Protestant Church to Ministry of the Interior, Berlin, October 10, 1935; memo-

randum, Preliminary Remarks, n.d.; BArch R1501/125221, letter, Institute for Business Cycle Research to Ministry of the Interior, December 13, 1933.

72. Erland Echlin, "Germany's Viewpoint," *Journal of Calendar Reform* 4, no. 2 (1934): 83–85, here 84.
73. LNA CTS, 1933–1946, 9a/6576/6576, Calendar Reform, Replies from Religious Authorities, League of Nations, Communications and Transit Section, Stabilisation des fêtes mobiles, August 3, 1934, Pan-Orthodox Congress; Universal Christian Council for Life and Work, International Christian Social Institute Research Department, The Churches and the Stabilisation of Easter: Results of the Enquiry into the Attitude of the Churches to the League of Nations Proposal (Geneva: International Christian Social Institute, 1933).
74. League of Nations Publication 4th C. G. C. T. 1, VIII. TRANSIT 1931. VIII. 12[11], Fourth General Conference on Communications and Transit, Preparatory Documents, vol. 2, Reply by the Holy See to the Invitation to Send a Representative to the Conference, Vatican to Secretariat League of Nations, Vatican City, October 8, 1931.
75. Carlyle B. Haynes, *Calendar Reform Threatens Religion* (Washington, DC: The Religious Liberty Association, n.d.), 1; see also BArch R/16/2198, memorandum on the conference of the German calendar reform committee on March 16, 1931, 11.
76. On the origins and religious connotations of the week, see Eviatar Zerubavel, *The Seven Day Circle: The History and Meaning of the Week* (New York: Free Press, 1985).
77. Joseph Hertz, *The Battle for the Sabbath at Geneva* (Oxford: Oxford University Press, 1932), 9; Moses B. Cotsworth, *Moses: The Greatest of Calendar Reformers* (Washington, DC: International Fixed Calendar League, n.d.); BArch NS/5/VI/17218, Resistance of the Rabbis against Changing the Calendar, Response of the International Fixed Calendar League from Director Moses B. Cotsworth to the elaborations by Chief Rabbi of the British Empire Dr. J. H. Hertz and other Rabbis and Opponents, 2. See also J. H. Hertz, *Changing the Calendar: Consequent Danger and Confusion* (London: Oxford University Press, 1931), 1.
78. LNA CTS, 1919–1927, 14/36440/12478, Calendar Reform, Advisory and Technical Committee, Replies from the religious authorities, Reply from the Chief Rabbi of the United Hebrew Congregation of the British Empire, London, May 30, 1924, 1; ibid., letter from the President of the Central Conference of American Rabbis, London, September 19, 1924, 2; LNA CTS, 1919–1927, 14/44863/12478, La réforme du Calendrier, Comité Special, Procès-verbal de la deuxième session, February 16, 1925, 1–4, listing five Jewish representatives in attendance who all made energetic statements; Ludwig Rosenthal, *"Juden der Welt erhaltet den Sabbat!" Referat zur Kalenderreform, auf dem Sabbat-Weltkongress am 2. August 1930 gehalten* (Berlin: Self-published, n.d.), 1; LNA CTS, 1928–1932, 9a/22806/1487, Correspondance avec le Comité Israélite concernant la réforme du calendrier, The Jewish Position on Calendar Reform, June 9, 1931, 2; see Rosenthal, *"Juden der Welt,"* 1.

79. LNA CTS, 1928–1932, 9a/31833/1487, Calendar Reform, Correspondence with the Jamiyat-e Ulama-e Burma, Rangoon, letter, Jamiyat-e Ulama-e Burma to Secretary General League of Nations, Rangoon, September 15, 1931.
80. LNA CTS, 1928–1932, 9a/35561/1487, Correspondence with Mr. Chowdhury, letter, G. A. Chowdhury to Secretary General League of Nations, Delhi, February 15, 1932.
81. Jean Nussbaum, "The Proposed Calendar Reform: A Threat to Religion," *British Advent Messenger* 60, no. 5 (March 4, 1955): 1–3; "International Organizations: Summary of Activities: Economic and Social Council," *International Organization* 10, no. 4 (1956): 618–627, here 618.
82. "Reform of the Hindu Calendar," *The Leader,* May 31, 1912, 6; on Parsi calendar reform, see "Parsi Festivals: The Unreformed Calendar," *Times of India,* August 13, 1915, 6.
83. LNA CTS, 1928–1932, 9a/1487/1487, Correspondence with Mr. Cotsworth, letter, Cotsworth to Robert Haas, January 26, 1931; LNA CTS, 1919–1927, 14/36440/12478, Replies from the religious authorities, Reply from the Confucian Association, Peking, China, Chen Huan-Chang, February 13, 1924.
84. LNA CTS, 1919–1927, 14/36384x/12478 Advisory and Technical Committee, Draft minutes of the second meeting of the committee of enquiry, May 20, 1924, 5, 8.
85. See Abigail Green and Vincent Viaene, "Introduction: Rethinking Religion and Globalization," in *Religious Internationals in the Modern World: Globalization and Faith Communities since 1750,* ed. Abigail Green and Vincent Viaene (Basingstoke, UK: Palgrave Macmillan, 2012), 1–19, here 7–9; Christopher A. Bayly, *The Birth of the Modern World, 1780–1914: Global Connections and Comparisons* (Malden, MA: Blackwell, 2004), 325; Gorman, *Emergence,* ch. 7.
86. LNA CTS, 1928–1932, 9a/25153/1487, Religious Liberty Association, various signature collections, September/October 1931; letter, Religious Liberty Association to Director of Communications and Transit Section, Poona, September 25, 1931; October 2, 1931; October 9, 1931; October 25, 1931. On opposition from the Seventh-Day Adventists, see also Seventh-Day Adventist Denomination, *A Petition to the League of Nations in Regard to the Revision of the Calendar* (Washington, DC: Self-published, 1931).
87. Stefan Zweig's famous autobiography (1942) describes life in the Austro-Hungarian Empire but can be read as a farewell to the nineteenth century more generally. Stefan Zweig, *The World of Yesterday: An Autobiography* (New York: Viking Press, 1943).
88. On the emergence of "the economy" as an analytical and colloquial concept, see Karabell, *Leading Indicators,* and Mitchell, *Rule of Experts,* 81–82.
89. Mark Mazower, *Governing the World: The History of an Idea* (New York: Penguin Press, 2012), 189.

## Conclusion

1. "Turning Their Backs on the World," *The Economist,* February 19, 2009, http://www.economist.com/node/13145370; "Going Backwards: The World Is

Less Connected than It Was in 2007," *The Economist,* December 22, 2012, http://www.economist.com/news/business/21568753-world-less-connected-it-was-2007-going-backwards; see also Dani Rodrik, *Has Globalization Gone Too Far?* (Washington, DC: Institute of International Economics, 1997).

2. Sven Beckert, *Empire of Cotton: A Global History* (New York: Knopf, 2014).
3. See O'Rourke and Williamson, *Globalization and History,* chs. 2, 3; see also Geyer and Bright, "Regimes," 218.
4. The two most encompassing contributions to recent attempts at writing global history illustrate the tension between similarity and difference: Bayly, *Birth of the Modern World,* and Osterhammel, *Transformation of the World.*
5. Paraphrasing Charles Tilly, *Big Structures, Large Processes, Huge Comparisons* (New York: Russell Sage Foundation, 1984).
6. AN F/17/2921, Commission de décimalisation, Pamphlet, Henri de Sarrauton, Deux projets de loi, 1.
7. On self-strengthening and targeted self-modernization in the non-Western world, see Philip Curtin, *The World and the West: The European Challenge and the Overseas Response in the Age of Empire* (Cambridge: Cambridge University Press, 2002); as one example, see Thongchai Winichakul, *Siam Mapped: A History of the Geo-Body of a Nation* (Honolulu: University of Hawai'i Press, 1994).
8. Roland Robertson, "Glocalization: Time-Space and Homogeneity-Heterogeneity," in *Global Modernities,* ed. Mike Featherstone, Scott Lash, and Roland Robertson (London: Sage, 1995), 25–44, and Arjun Appadurai, *Modernity at Large: Cultural Dimensions of Globalization* (Minneapolis: University of Minnesota Press, 1996).
9. On the emergence of holistic civilizational identities and pan-movements, see Cemil Aydin, *The Politics of Anti-Westernism in Asia: Visions of World Order in Pan-Islamic and Pan-Asian Thought* (New York: Columbia University Press, 2007).
10. On the discovery of populations, see Michel Foucault's concepts of biopolitics and governmentality: Foucault, *The Birth of Biopolitics: Lectures at the Collège de France, 1978–1979,* ed. Michel Senellart (London: Palgrave Macmillan, 2008); Foucault, "Governmentality," in Foucault, *Security, Territory, Population: Lectures at the College de France, 1977–78* (London: Palgrave MacMillan, 2007), 126–145; Charles S. Maier, "Leviathan 2.0: Inventing Modern Statehood," in Rosenberg, *A World Connecting,* 29–283, here 156–170. In the amended second edition of his *Imagined Communities,* Benedict Anderson added a chapter on maps and censuses that echoes some of these arguments. Anderson, *Imagined Communities,* ch. 10.
11. Torp, *Herausforderung,* esp. 369–370.
12. Hassan Kayalı, *Arabs and Young Turks: Ottomanism, Arabism, and Islamism in the Ottoman Empire, 1908–1918* (Berkeley: University of California Press, 1997).
13. Osterhammel, *Transformation of the World,* 29–35.
14. On early paper and print from a Middle Eastern perspective, see Nile Green, "Journeymen, Middlemen: Travel, Trans-Culture and Technology in the

Origins of Muslim Printing," *International Journal of Middle East Studies* 41, no. 2 (2009): 203–224; Isabella Löhr, *Die Globalisierung geistiger Eigentumsrechte: Neue Strukturen internationaler Zusammenarbeit, 1886–1952* (Göttingen: Vandenhoeck & Ruprecht, 2010).

15. Thompson, "Time, Work Discipline, and Industrial Capitalism," 58–59.
16. Anderson, *Imagined Communities*, 36.

# Acknowledgments

As I write these words at the local Starbucks in Kew Gardens, United Kingdom, while suffering from a particularly bad bout of jet lag, the limits of time standardization are tiresomely clear. Our time zones may be entirely uniform, and without time management we would be lost in our busy lives. But until the marvels of modern medicine will produce a substitute for sleep and rest, human bodies and minds will always revolt against the rapidity with which oceans and continents can be crossed today thanks to a process of time coordination that underlies modern transportation but began in the second half of the nineteenth century—which is the story of this book.

The greatest moments of archival delight on the journey to finishing this book always entailed bits and pieces of evidence showing how much hard work went into making time more uniform, and how "unnatural" this process by all means was. I have to thank a number of teachers and colleagues for teaching me how to notice and read these moments, and how to think—and not to think—about time. This book began at the Free University of Berlin and Harvard University. At Harvard, I had the great fortune of working with Sven Beckert and Charles S. Maier. Sven's enthusiasm for genuinely global histories meant that I was able to pursue a project that others would have turned down as too vast and simply unmanageable. Sven also asked questions about the relationship between capitalism and globalization and the role of the state in globalization, matters that continue to preoccupy me even as I embark on a new project on a completely different topic. Charlie, as he is known at Harvard, brought a necessary skepticism (and humor) to every topic and prodded me to think big.

Others at Harvard have been equally supportive and important. Afsaneh Najmabadi was willing to work with me at a point when the project looked as if it would evolve more in a Middle Eastern than a global direction. She stayed on despite the change of course, and I am profoundly grateful for that. Roger Owen, David Armitage, and Erez Manela were partners in conversation or conveners of seminars that helped me think about global history. Two terrific Arabic teachers,

Carl Sharif Eltobgui and Khaled al-Masri, motivated me to mobilize the effort it requires to learn this language.

My colleagues in my new department at the University of Pennsylvania, where I have been teaching since 2011, soon made College Hall feel like my academic home. In particular, I have to thank Warren Breckman, Steve Hahn, Peter Holquist, Benjamin Nathans, Amy Offner, Tom Sugrue, Eve Troutt-Powell, and above all, Stephanie McCurry, for offering advice, comments on written work, and much more to a new colleague. I could not have completed this book without the generous funding, marvelous staff, and amazing collegial company at the Institute for Advanced Study's (IAS) School of Social Sciences, where I spent a year on academic leave. I thank Danielle Allen, Didier Fassin, Dani Rodrik, and Joan W. Scott for hosting me, and our fantastic group of colleagues in 2013–2014 for making this one of the most memorable experiences of my academic life so far.

Finishing this book has taken me to a wide range of archives and sites, where various people helped out over the years. I want to mention the staff at the American University of Beirut's Jafet Library, the director and staff at the German Orient Institute in Beirut, and above all the wonderful group of librarians at the IAS library.

Several people helped me out with research assistance, translations, and other tasks and logistics over the years—Cameron Cross, Werner Hillebrecht, Jonathan Korn, Nathaniel Miller, Thanh Nguyen, Andrea Okorley, Emma Paunceforth, Roberto Saba, and Philippa Söldenwagner.

This book required financial support offered generously by various sources of funding: the Center for European Studies and the Graduate School of Arts and Sciences at Harvard; the Studienstiftung, the German Academic Exchange Service/ DAAD; the Smithsonian Institution's National Museum of American History; the German Orient Institute in Beirut; the University of Pennsylvania Research Foundation and the School of Arts and Sciences at Penn, among several others.

Many colleagues and friends have over the years lent an eye or ear to provide feedback on my work and have invited me to present in their seminars and shared published and unpublished materials with me: in particular, David Armitage, Orit Bashkin, David Bell, Mark Bradley, Shel Garon, Michael Geyer, Rob Kohler, Noam Maggor, Erez Manela, Adam McKeown, Kathryn M. Olesko, Aviel Roshwald, Jacob Soll, James Vernon, Jakob Vogel, Tara Zahra, and the participants in the history of capitalism seminar at Harvard. Sebastian Conrad's work has been a constant source of inspiration and admiration for my own thinking about history ever since my time as a student in Berlin. Till Grallert graciously pointed out errors to me; all remaining ones are, of course, my own.

At Harvard University Press, Joyce Seltzer took an early interest in my ideas when they barely constituted a book project. Brian Distelberg assured easy and swift communication and helped with many other things. The outside reviewers for the press provided essential and creative feedback that permitted me to complete a better book.

Some of the ideas discussed in Chapters 4 and 5 were explored previously in "Whose Time Is It? The Pluralization of Time and the Global Condition, 1870s–1940s," *American Historical Review* 120, no. 5 (December 2013): 1376–1402.

The most important support in writing this book came from all those who offered their friendship and emotional sustenance over the past years, whether in Cambridge, Massachusetts; Beirut; Philadelphia; or places in between—they know who they are. To them and to my family I owe the greatest debts—debts that cannot be repaid. In some ways this book is for J., who wouldn't know but probably taught me most about human time.

# Index